LIVING WITHOUT LANDFILLS

LIVING WITHOUT LANDFILLS

Confronting the "Low-Level" Radioactive Waste Crisis

By Marvin Resnikoff, Ph.D.
Edited by Steven Becker
Design and Production by Ed Hedemann

A Special Report by the Radioactive Waste Campaign

New York, New York

Commercial "Low-Level" Waste Project Advisory Board

Sr. Rosalie Bertell, Ph.D., Institute of Concern for Public Health
Dee D'Arrigo, Nuclear Information and Resource Service
Kay Drey, Coalition for the Environment
Maarten DeKadt, Ph.D., INFORM
*Lisa Finaldi, Clean Water Fund
Minard Hamilton, Radioactive Waste Campaign
James Goldstein, ESRG
Joanna Hoelscher, Citizens for a Better Environment
Judith Johnsrud, Ph.D., Environmental Coalition on Nuclear Power
Charles Komanoff, Komanoff Energy Associates
Robert K. McLellan, M.D., M.P.H., Gesell Institute of Human Development
Ellen Messing, Esq., Stern & Shapiro
*Carol Mongerson, Coalition on West Valley Nuclear Wastes
Warren P. Murphy, Vice President, Vermont Yankee
*June Peoples, editor
Robert Pohl, Ph.D., Cornell University
Jeff Schmidt, Sierra Club
Ernst Schori, Dartmouth College
Mary Sinclair, Great Lakes Energy Alliance
Peter Skinner, P.E., New York Attorney General's Office
Gordon Thompson, Ph.D., Institute for Resource & Security Studies

* *Member of the Board of Directors, Radioactive Waste Campaign*

The Advisory Board reviewed and commented upon the outline and table of contents of the study, and on a draft version of the study. Though not officially part of the Advisory Board, the Nuclear Regulatory Commission also reviewed and commented upon a draft of the study. The author and the Campaign take full responsibility for the accuracy of the study's research and the validity of its conclusions and recommendations. Organizational affiliations of Advisory Board members are provided for identification purposes only.

Copyright © 1987 by the Radioactive Waste Campaign, Inc.
All rights reserved
Published by Radioactive Waste Campaign
625 Broadway, 2nd Floor
New York, New York 10012
Library of Congress Card Catalog Number: 87-62278
International Standard Book Number: 0-9619078-0-0
Printed by Faculty Press, Brooklyn, New York 11218
Typeset by Mar + x Myles, New York, New York 10016
Your Type, New York, New York 10011
Designed by Ed Hedemann
Cover design by Rick Bickhart
  Photo by Ed Hedemann
Indexed by Elliot Linzer
Printed on 70 percent recycled paper

Table of Contents

Acknowledgements vii
Preface ix
INTRODUCTION 3
   Nuclear Mythology 4
   The Plan of the Book 5
CHAPTER 1·   WHAT IS "LOW-LEVEL" WASTE 9
   Nuclear Power Reactors 10
   Institutional Waste 18
   Industrial Waste 19
   Hazardous Life 24
   The Big Picture 26
CHAPTER 2   EXPERIENCE AT "LOW-LEVEL" WASTE LANDFILLS 33
   Maxey Flats 35
   Sheffield 36
   West Valley 37
   Barnwell 40
   Richland/Beatty 41
   The Verdict on Landfills 41
CHAPTER 3   THE GOVERNMENT RESPONSE: *TOO LITTLE, TOO LATE* 45
   New NRC Regulations 45
   NRC Scenarios Unrealistic 50
   NRC Adopts Weakened Regulations 51
   NRC Defines the Problem Away 51
   It's Magic: HLW Becomes LLW 52
   Federal Legislation--Congress Drops the Hot Potato 53
   The 1985 Amendments 53
CHAPTER 4   ALTERNATIVES TO RADIOACTIVE LANDFILLS 59
   Conservation 59
   Volume Reduction 60
   Separation of Waste by Half-Life 62
   Storage/Disposal 64
   Above Ground Systems 64
   Below Ground 70
   Economics of Waste Management 72
CHAPTER 5   CONCLUSIONS AND RECOMMENDATIONS 77
APPENDICES 81
   Glossary 83
   Review Comments by NRC and Response by RWC 87
   Contacts 89
References 97
Index 103

List of Tables

Table 1-1   Hazardous Life of Representative "Low-Level" Waste   13
Table 3-1   Change from Proposed to Final 10 CFR Part 61 Regulations   46
Table 3-2   Bone Dose Due to Group 4 Waste Streams in the Intruder/Discoverer Scenario   47
Table 3-3   Classification of PWR Decommissioning Waste   48
Table 3-4   10 CFR Part 61 Classification of Decommissioned Reactor Waste   49
Table 4-1   Unit Disposal Cost Estimates   73

List of Figures

Figure 1-1   Boiling Water Reactor Vessel and Internals   10
Figure 1-2   PWR "Low-Level" Waste Without Irradiated Components (curies)   11
Figure 1-3   PWR "Low-Level" Waste Without Irradiated Components (volume)   12
Figure 1-4   PWR "Low-Level" Waste (volume)   14
Figure 1-5   PWR "Low-Level" Waste (total radioactivity generated)   14
Figure 1-6   Total BWR Waste (volume)   15
Figure 1-7   Total BWR Waste (curies)   16
Figure 1-8   Total PWR Waste (radioactivity as a function of time)   17
Figure 1-9   Institutional "Low-Level" Waste (curies generated from 1980 to 2020)   18
Figure 1-10   Institutional "Low-Level" Waste (volume generated from 1980 to 2020)   19
Figure 1-11   Industrial LLW Generators (curies initially generated)   20
Figure 1-12   Tritium Manufacturers (curies)   20
Figure 1-13   Tritium Manufacturers (volume)   21
Figure 1-14   Industrial LLW Generators (curies after 100 years)   21
Figure 1-15   Industrial LLW Generators (volume)   22
Figure 1-16   Annual Commercial LLW Radioactivity   23
Figure 1-17   Annual Commercial LLW Volume   24
Figure 1-18   "Low-Level" Waste Without Decommissioned Waste (curies initially generated)   25
Figure 1-19   "Low-Level" Waste Without Decommissioned Waste (curies after 100 years)   26
Figure 1-20   "Low-Level" Waste Including Decommissioned Waste (curies initially generated)   27
Figure 1-21   "Low-Level" Waste Including Decommissioned Waste (volume)   28
Figure 1-22   "Low-Level" Waste Including Decommissioned Waste (radioactivity as a function of time)   29
Figure 2-1   Location of Waste Sites   34
Figure 3-1   "Low-Level" Radioactive Waste Compacts   54
Figure 4-1   Controlled Combustion Incinerator   62
Figure 4-2   "Low-Level" Waste   63
Figure 4-3   Storage Building   65
Figure 4-4   Above Ground Vault   66
Figure 4-5   Quadricells   67
Figure 4-6   Westinghouse Surepak System   68
Figure 4-7   Earth Mounded Concrete Bunker   70
Figure 4-8   Augered Hole   71

Acknowledgements

This book was written for environmental activists and state and local officials. It therefore could not be, and was not, written by one person alone sitting in a quiet room in New York City, divorced from the grass roots movement. Citizens threatened by a waste dump or incinerator vitally need specific information, "the facts." Determining which facts is an iterative process, from the author to the field, from the field to the author. In an important sense, I served as an assembler of collective wisdom. That is, this book has many authors. Only the largest contributors will be mentioned here. Please excuse me if someone has been overlooked.

To start with, the work of the Board of the Radioactive Waste Campaign should be acknowledged. Unlike boards of other organizations whose members are selected as big givers or big names while the staff runs the show, the Campaign's Board is composed of 15 radioactive waste activists. The staff does not run this organization. All projects are approved by the Board, which has a major role in the subject matter and policy recommendations. Special thanks to a Board Subcommittee composed of Lisa Finaldi, Carol Mongerson and June Peoples, who participated in detailed editing and writing of the book. Chapter 5 and the policy implications were discussed by the entire Board. Thanks also to Board member Warren Liebold for valuable suggestions on a later draft of the manuscript. Each Board member has many other responsibilities—a job, a family, a home, and a life—yet, each gave many hours and weeks to thinking and writing about radioactive waste.

Many thanks to Minard Hamilton, our Director, for her valuable comments and suggestions. Her recommendations led to a much more comprehensive policy, as enunciated in Chapter 5. Servicing the grass roots movement, state and local legislatures, and the media, is a major effort, involving a hidden, but real, administrative nightmare, from the most important policy matters to the most trivial (pay checks, rent, stamps, phones, desks, stationery, on and on). Who communicates with foundations and the Board of Directors? Who coordinates policy and direction? Who writes editorials for *the Waste Paper?* As the Campaign has grown, so has the administrative burden. Without our Director, Minard Hamilton, taking on many of these thankless tasks, my time would never have been freed up to complete this work.

Similarly, Jennie Tichenor, our Office Manager, has taken on the lion's share of communicating directly with the grass roots movement and responding to information requests. Without her help, I would have been talking on the phone all day. Thanks also for her sense of humor and moral support. Jennie also compiled the Contacts section in the Appendix, as well as taking care of the financial records and writing for *the Waste Paper.*

To be able to produce this book, money is needed. Proceeds from sales can hardly pay the tab. We are thankful to many foundations and individual contributors who believed enough in our work to assist us. In particular, we gratefully thank the anonymous contributors (you know who you are!) and following foundations for their assistance: Conservation and Research Foundation, Fund for New Jersey, New-Land Foundation, New York Community Trust, Public Welfare Foundation, and Rockefeller Family Associates.

While the entire Review Panel (see page iv) was tremendously helpful, I'd like to single out the heroic work of the following members, some of whom commented upon almost every page of the draft: Sr. Rosalie Bertell (International Institute of Concern for Public Health), Dee D'Arrigo (NIRS), Marty DeKadt (INFORM), Kay Drey (Missouri Council on the Environment), Lisa Finaldi (Clean Water Fund and Chair, Radioactive Waste Campaign), Minard Hamilton (Director, Radioactive Waste Campaign), Joanna Hoelscher (Citizens for a Better Environment), Carol Mongerson (Coalition on West Valley Nuclear Wastes and Vice-Chair, Radioactive Waste Campaign), and Ernst Schori (Dartmouth College). Almost all comments and suggestions filled in important omissions and were incorporated.

Also helpful was the Nuclear Regulatory Commission. Though the book is based upon their data, they will not be pleased with the analysis, conclusions and recommendations. This is to be expected; we listen to different drummers. The NRC is closer to waste generators; the Campaign is closer to real people affected by waste dumps. The NRC's comments and our responses are included in an Appendix.

This book began under the aegis of the Sierra Club. We gratefully acknowledge many enlightening discussions with Ellen Winchester, Jesse Riley, Warren Liebold, Jack Neff, and the many grass roots volunteers who give lifeblood to that organization. We interacted with Sierra Club headquarters through staffperson Gene Coan and greatly acknowledge his encouragement and good humor as we jointly put out the many brush fires that arose.

Once all the comments have been incorporated and the information is down on paper, can anyone understand it? Is the book well-organized, consistent and readable? Thanks here go to the editor, Steven Becker, who has given much detailed and careful thought to organization and policy implications. Steve worked assiduously on everything from sentence structure to major strategic thinking. His efforts were extremely valuable.

The work of Dave Pyles and Bill McDonnell, for fact-checking and glossary compilation is gratefully acknowledged.

Once all the words are written and edited, and the tables and figures assembled, then how does it become a book? Enter Ed Hedemann who did the design work, specked type, did the layout, and worked with the typesetter and printer, Faculty Press. Without this production work, there is no book. Ed often worked 60 hours per week, chasing one detail after another.

If this book serves to enlighten citizens and state and local officials about the nature and hazard of "low-level" waste, and how to protect ourselves, then our team has done our part of the job. With this book as a guide, the next steps are yours. Because of federal and state legislation, the time is now. Get in touch with us and let us know how we can better assist you in your work. Use the contact list in the Appendix to get in touch with a local group. Citizens working together can make all the difference.

Marvin Resnikoff
July 20, 1987

Preface

Nine years ago Lois Gibbs mobilized citizens to protest a high incidence of miscarriages, birth defects and other health effects at Love Canal. The media jumped on the story and overnight toxic chemicals and the threat they pose entered the public consciousness. What Gibbs did for the issue of toxic chemical pollution echoed what Rachel Carson had done for pesticide contamination with the publication of *Silent Spring* in 1962.

Living Without Landfills has the potential to play the same key role for a hitherto neglected segment of the nuclear garbage dilemma—"low-level" waste. In part because of misleading nomenclature, in part because of a very successful nuclear industry public relations campaign, citizens, the media and legislators have ignored this vital issue.

After all, if we are only talking about a few mildly contaminated plastic gloves, booties, tools and other materials that will be hazardous for a short period of time, whey get upset? In particular, let's not get upset if our doing so delays the siting of new dumps and puts in jeopardy medical care for any loved ones who may be suffering from cancer. Let's just dump the essentially benign refuse in an "engineered" landfill, guard the dump for a few decades, then convert it to a golf course.

Living Without Landfills slips the mask off of this industry-sponsored myth. As nuclear physicist and Campaign Research Director Dr. Marvin Resnikoff details in this book, "low-level" waste is a misnomer. "Low-level" waste includes extremely long-lived, intensely radioactive waste as well. Materials that will be hazardous for thousands and thousands and thousands of years are, currently, being irresponsibly dumped in "low-level" burial grounds. These materials include such contaminants as iodine–129, niobium–94 and nickel–59. They must immediately be removed from the "low-level" category and banned from landfills.

Further, by far the highest contribution to "low-level" waste both by volume and toxicity comes from commercial nuclear power plants. A staggering 99 percent of the radioactivity in "low-level" waste comes from nuclear reactors. And most of the *longest-lived* nuclear contaminants are generated by reactors. These facts contrast sharply with utility innuendo which has suggested "low-level" waste is generated mainly by medical or research activities.

Living Without Landfills cogently puts the facts out on the table. Thus far, the lack of facts has impeded the development of safe and sane approaches to managing "low-level" waste. By no means, however, are there easy answers to this intractable problem. When considering what to do with toxins that will be poisonous for as many centuries forward into human history as Cro-Magnon women and men are back, is, to put it mildly, mind-boggling.

This is one of several reasons the Radioactive Waste Campaign is advocating the launching of Manhattan Project II. Manhattan Project I was initiated during World War II. The U.S. government, galvanished by the threat that German scientists would come up with the design for the A-bomb first, sent the best scientific minds of the time into the desert with an almost unlimited budget and a research agenda free of bureaucratic strings.

The radioactive waste crisis we face in 1987 requires equally drastic action. Despite bland assurances by government and industry, we are threatened with the potential permanent contamination of underground water supply systems, the poisoning of significant agricultural resources, and the assured deadening of vast tracts of land. We are also risking the health of future generations.

Manhattan Project I was, of necessity, a top-secret wartime program. Manhattan Project II must, of necessity, be an open program. All proceedings of Manhattan Project II should be laid bare to public scrutiny and an open spirit must prevail.

Some may argue that the launching of a scientific program of this magnitude is dangerous—it will delay a solution to the nuclear waste problem. We say that when the health and safety of future generations is at risk, a delay is preferable to an irresponsible course of action. Given the stakes, a delay is highly ethical and appropriate.

As readers will note, the Radioactive Waste Campaign supports an orderly, rational phase-out of nuclear reactors. While Manhattan Project II is underway, production of nuclear wastes must be minimized.

For too long, the safe energy movement has operated within a climate of fear. Lobbyists, particularly in Washington, have argued we cannot support phase-out, or our Capital Hill credibility will suffer. Unfortunately, the grass roots often have been mesmerized by this timid approach. Perhaps the argument was valid in the early conservative bloom of the Reagan years. Now, however, our credibility is suffering by not honestly confronting the need to convert to conservation and renewable energy strategies.

As we convert to safe energy sources, careful attention must be paid to the needs of displaced workers. Any shift to renewable energies must involve the allocation of monies for job retraining and job placement of ex-nuclear workers.

Our support for phase-out is not just because of the legacy of hazardous-for-eons nuclear garbage that is churned out by reactors on a daily basis. We are concerned about the impact of nuclear technology on democracy. In the "low-level" waste arena, regional commissions directed by political appointees are now empowered to site new nuclear dumps. This is a troubling step away from accountability and democratic decision-making.

The best protection against an erosion of democracy is an informed and empowered citizenry. From the outset, the Radioactive Waste Campaign, founded in 1978 in New York, has considered it our mandate to provide the public with rigorously accurate, scientifically impeccable information that is also accessible. It has also been our goal to empower citizens, to help them use our scientific materials to both educate themselves and to organize neighbors, communities, legislators and the media.

It is in a spirit of hopefulness for a better world for our children and grandchildren that we publish *Living Without Landfills*.

Minard Hamilton
July 20, 1987

Introduction

Introduction

The "low-level" waste issue is hot. Battle lines are forming across the country. As the federally-mandated deadline for choosing states to host waste facilities draws near, state legislatures, spurred on by citizen activism, have become edgy. Even when states are not selected, but just in the running, citizen concern has become a powerful force.

In the central part of the country, for example, 109 of the 148 sites being considered as possible locations for a waste facility are in Kansas. The news was enough to have North Central Kansas Citizens turn out 6,000 angry people at a March 1987 hearing in the town of Beloit (population 4,300). And at a hearing later that month before the Kansas Senate Subcommittee on Energy and Natural Resources, the group's president, 36-year-old housewife Laura Menhusen, did not mince words: "This will be the most important issue you will ever face in your lifetime." Menhusen had never before been involved in political activity (Russell,1987).

In New Jersey, the Department of Environmental Protection proposed that earth contaminated with radium and other long-lived radionuclides be dug from near people's homes in Montclair and transferred to Vernon, near the New York border. The town of Vernon has a population of less than 500, but a summer rally to oppose the planned transfer drew 5,000 friends and neighbors. Protests in the state capital, Trenton, and at the governor's mansion, drew 1,000 citizens each. Mild-mannered, law-abiding citizens were ready to lie down in the streets to prevent their neighborhoods from being defiled.

As a result, the Department of Environmental Protection withdrew its plans and set up an advisory committee to determine where about *200 curies* of contaminated dirt should go. Citizens are urging the state to store the radium soil at McGuire Air Force Base, on a concrete pad where a 1963 Bomarc missile explosion spewed plutonium over an 11-acre area. The state has so far refused to consider this option, even though it is likely no community will accept this radioactive waste. The question arises—if New Jersey cannot find a home for 200 curies, how will it (or any other state) find a place for many times that amount of "low-level" waste?

Long neglected by the nuclear industry as lacking "glamour," the radioactive waste issue is now on center stage. And citizens are making a difference. The huge quantities of nuclear garbage that have accumulated across the country, and the past failures by industry and government to properly address the problem, have left us as a society with no choice but to come to terms with this issue now. Radioactive waste has become the subject of legislation, regulation, hearings and litigation. It has sparked rallies, protests and civil disobedience. For some people it is big business. For many more it is a cause for great concern.

This book examines a central part of the nuclear waste problem: commercial "low-level" radioactive waste. A detailed look at this topic is perhaps more timely than ever before. Under legislation passed in 1980 and amended five years later, all 50 states are required, either individually or in regional associations, to develop facilities for managing all the commercial "low-level" waste generated within their borders.

With deadlines just around the corner, legislators and policymakers are in high gear to come up with plans and put them into place. Meanwhile, citizens are expressing grave reservations about the wisdom of some of those plans. Indeed, the stakes involved are awesome: choices made today could have ramifications for the next 100,000 years, the time it takes for some long-lived isotopes in "low-level" waste to decay.

Just what is "low-level" waste? How does it arise? Who are the generators? How long-lived and hazardous is "low-level" waste? How has this waste been dealt with in the past? And, what should be done with radioactive waste in the future? These are the vital questions explored in great detail in this book. Our ultimate aim is to provide guidelines for a sensible, environmentally-sound, safety-first approach to managing "low-level" radioactive waste.

Nuclear Mythology

Comprehensive discussions of the "low-level" radioactive waste issue are difficult to find. Perhaps part of this stems from the name "low-level." It somehow suggests, misleadingly, that the waste is not important, not dangerous. Whatever the case, the nuclear industry has been more than happy to step into the void and fill the information gap with its own self-serving mythology.

In the picture the nuclear industry has painted, there are 20,000 generators of waste providing a multitude of beneficial uses of radioactivity. And from this picture comes an inescapable conclusion: without an immediate solution to the waste disposal problem, cancer patients would lose treatment, and products which "enrich our lives" would be lost. As the Atomic Industrial Forum has warned ominously (see box), all the beneficial uses of radioactivity—smoke detectors, jet engine inspection, patient diagnosis, perfume, paper and baby formula manufacture—would disappear (AIF,1986).

Taking up the cause, the New York Voice of Energy ("Pro Energy, Pro Growth and Pro America"), a nuclear power advocacy group, has argued, "In terms of human health and safety, no delay or disruption of lifesaving procedures can be tolerated! Some...physicians and educators will ultimately face curtailment of their ability to provide treatment, laboratory procedures or effective teaching methods (Voice,1984)." Going still further, the New York State Low-Level Waste Group claimed that 500,000 jobs were on the line unless a "low-level" waste facility was in operation by December 1985 (LLWG,1985).

NUCLEAR INDUSTRY MYTHOLOGY:
Presenting Nuclear Power Plants As Just One Waste Generator Among Many

Awakened by smoke alarms, a family flees into the night to escape a house in flames. A jet airliner gracefully lifts off the runway, minutes after its engines passed inspection for internal damage. While the patient watches, an animated image of his suspect heart appears on a color video screen. Electricity flows from a nuclear power plant to energize homes, schools, stores, offices and factories in a city of 500,000 people. Preparing for an evening out, a young woman rubs a spot of perfume on each wrist and behind each knee. A mile-long sheet of paper takes on a coating of clay and titanium in just over a minute to make the glossy pages for magazines and booklets. Bottle of formula at hand, a contented infant settles back for an afternoon snack and snooze.

Atomic Industrial Forum
(AIF, 1986)

Doesn't it appear odd that such a dire warning about a loss of quality of life is coming from the nuclear power industry? It is more than a public service. When the basic facts of "low-level" waste are understood, then the AIF "concern" becomes crystal clear. The nuclear industry wants to draw public attention away from the fact that it is by far the largest generator of commercial radioactivity. Taking into account all commercial "low-level" radioactive waste, 100 nuclear reactors produce about 99 percent of the radioactivity in "low-level" waste in the United States. Take all the uses of radioactivity from the rest of the 20,000 waste generators, add them up, and you will not get more than 1 percent of the radioactivity.

Unfortunately, the mythology developed by the nuclear power industry has been spread far and wide. Descriptions of "low-level" waste as "contaminated refuse," as just "mildly radioactive," or as a "more benign" variety of nuclear refuse are commonly found in the media (Carlson,1987).

Nuclear industry mythology also permeates public statements by waste generators who hardly produce any radioactive waste. Medical practitioners state that "low-level" radioactive wastes are "produced by myriad commercial activities—industrial, medical, and research—as well as through the production of electricity by nuclear energy (ACP,1984)." The American College of Physicians even believes that all "low-level" waste is "characterized by relatively low levels of radioactivity, generally requiring little shielding and no cooling, and usually representing a potential hazard of limited duration (generally a half-life of hours or days)." This may be true for medical waste, but certainly not for nuclear power reactor waste. Radioactivity emitted from some radioactive stainless steel components can produce lethal radiation doses in less than a minute, surely requiring more than a "little shielding." Internal components can stay radioactive for over 100,000 years, hardly a "limited duration" in any meaningful sense.

The dire warnings from industry, then, are hiding the basic facts. The medical community is being used as a lever to open new waste facilities for the nuclear power industry. In football talk, the institutional and industrial waste generators are being used to "run interference" for the nuclear industry. Since nuclear reactors, with arguable benefits, produce so much radioactivity, of course the industry hides behind hospital and university generators, less controversial uses of radioactivity. Institutional waste is only the pimple on an elephant. Clearly, medical practitioners (and the rest of us) need to pierce this mythological veil and become conscious of the basic facts of "low-level" waste.

Though great richness of detail will be provided in this book, the basic 99 percent vs. 1 percent ratio should be kept in mind. It can be understood in the following way. Cintichem, with a 5-megawatt(t) reactor, produces 50 percent of the radioisotopes in medical use in the United States. Contrast the Cintichem reactor with a large nuclear power reactor having a 3,500-megawatt(t) power rating, multiply this by 115 to account for all commercial power reactors in the United States, and you get a ratio of 70,000 to 1, nuclear power reactor waste to radioisotope waste. Admittedly this is a broad brush generalization, but it holds up under closer scrutiny.

The Plan of the Book

We start with a multi-faceted look at the basic facts of "low-level" waste. In Chapter 1, we discuss in detail what the term "low-level" means, how "low-level" waste is created, and who generates it. We present a detailed discussion, waste stream by waste stream, of the volume, radioactive content and short and long-term hazard of each type of "low-level" waste. Though we point out the major producers of radioactive waste in various regions, our discussion deals with waste generated in the country as a whole.

We have sought to identify all the types of radioactive waste entering commercial waste facilities. This includes a wide range of materials. "Low-level" waste is hardly a homogeneous category—it includes just about everything that is not considered "high-level" waste (irradiated fuel). This leaves a tremendous spread in radioactive hazard in different waste streams. Some waste streams are extremely radioactive and long-lived, while others, particularly from medical and research institutions, are only mildly radioactive and short-lived.

To research the individual waste streams, we relied primarily on one information source, a Nuclear Regulatory Commission report, *Update of Part 61 Impacts Analysis Methodology*, produced by the Envirosphere Company (NRC,1986a). This comprehensive report identifies 256 significant waste streams, each containing, on average, 10 different long-lived radionuclides.

With this data, the NRC contractor, unfortunately, did not do the basic comparisons and analysis we perform here. By comparing waste streams and evaluating their hazard over short and long time frames (over 10,000 years), we could identify the major and minor actors. Who are the major producers of "low-level" waste? Which waste streams have high concentrations of long-lived radionuclides? Which have low concentrations of short-lived radionuclides? To answer these questions and more, over 25,000 data points were calculated. They are summarized in charts and graphs in the pages that follow.

We did not investigate waste streams which go into Department of Energy landfills. For example, in the production of plutonium for nuclear weapons, the Energy Department annually creates a volume of waste approximately equal to that generated by the commercial sector (DOE,1986). This waste contains about twice as much radioactivity as commercial waste.

Further, the widespread distribution of military high-level waste makes commercial "low-level" waste. Tritium produced by the Energy Department for nuclear weapons is distributed to industrial suppliers, who, in turn, make it available to universities and institutions. The fact that all tritium used by hospitals and research institutions comes as a by-product of nuclear weapons production is rarely mentioned at our institutions of higher learning.

Many other radioisotopes, such as americium and cesium, are provided by Oak Ridge National Laboratory, and also come from nuclear weapons production. In a sense, there is a little bit of a bomb in each smoke detector. Americium-241 is extracted from the plutonium in older nuclear warheads at the Rocky Flats Plant. And cesium, used in food irradiation, comes from high-level radioactive waste itself. We do not consider the policy and health implications of these practices in this book. However, some of these topics will be explored in the *Citizen's Guide to Military Landfills*, the Radioactive Waste Campaign's forthcoming book on military nuclear waste facilities.

In Chapter 2, we present a brief history of radioactive landfill experience in the United States. Included are discussions of specific waste sites, as well as an overall assessment of landfills.

Chapter 3 looks at the government's response to the "low-level" waste problem. Waste management regulations adopted by the Nuclear Regulatory Commission, 10 CFR Part 61, are carefully explained and assessed. Although these regulations are an improvement over past practices, we are not reassured they will lead to safe waste management facilities. Chapter 3 also includes a discussion of recent efforts to redefine some waste as "below regulatory concern," and an examination of the Low-Level Radioactive Waste Policy Act.

In Chapter 4, waste management alternatives are described and matched to waste streams. Ways to effectively isolate small-volume, high-concentration waste streams are identified, as are methods for dealing with less long-lived streams.

Conclusions and recommendations appear in Chapter 5. Comments from the Nuclear Regulatory Commission and our response follows Chapter 5, as does a glossary and an appendix listing local environmental organizations active on the "low-level" waste issue.

While this book will clear away many myths about radioactive waste, and provide guidance on waste management, do not look for a Hollywood ending here. There can be no such simple answer to this complex problem, and it would be irresponsible for anyone to say there can be. Much of the generated waste is long-lived and will remain hazardous for hundreds of years. Some waste streams will even be hazardous after tens of thousands of years.

Very surprisingly, though, we do find that the long-lived portion of "low-level" waste occupies a very small volume, which facilitates its isolation. Further, a large volume of "low-level" radioactive waste poses only a short-term hazard, that is, a 100-year problem. These materials can be reduced in volume and stored until they decay. We also explore the important policy implications of the fact that 99 percent of total "low-level" waste radioactivity and an even greater percentage of the longest-lived radioactivity is generated at nuclear power plants.

None of the management options we recommend involves landfills. This means they are more expensive in the short term than present proposals. Among those proposals, landfills are, of course, the least expensive from the perspective of waste generators, since long-term expenses, which have been high, are borne by taxpayers. Undoubtedly, some people in industry and government will ignore past failures and want to "dispose of" radioactive waste in unlined trenches. In this book, though, we will show that it is possible to live without landfills, and we will discuss alternatives which are far more sensible from the standpoint of safeguarding the environment and public health, now and in the future.

CHAPTER 1

What Is "Low-Level" Waste?

CHAPTER 1

What Is "Low-Level" Waste?

In the United States, "low-level" waste is what it isn't—that is, it is defined as all waste that doesn't fall into any of the other established categories of radioactive waste. "Low-level" waste is everything that is not high-level waste, not uranium tailings, and not transuranic waste.* Obviously, then, the term "low-level" waste covers a broad spectrum of radioactive materials. These range from slightly contaminated booties and test tubes to intensely radioactive metals drawn from the interior of a nuclear reactor. The hazardous life of "low-level" waste ranges from hours to hundreds of thousands of years.

"Low-level" radioactive waste is produced by three major categories of generators: industry, institutions (medical and research) and nuclear power reactors. Industry produces and employs radioisotopes for such uses as well-logging, radiography, food processing and luminescent signs. Radioisotopes are used in medicine for diagnostic studies of organs, radioimmunoassay, and so on, as well as in medical and agricultural research. And in nuclear reactors, "low-level" wastes are produced in the course of generating electricity.

All told, throughout the U.S., over 20,000 licensees handle radioactive materials and generate radioactive waste. Nevertheless, as we show in this chapter, when *all* commercial "low-level" waste is accounted for, a small number of commercial generators, nuclear power reactors, produce a whopping 99 percent of the radioactivity.

How have these wastes been managed in the past, and how might they be dealt with in the future? Before we can begin focusing on these issues, we need to more closely examine this amorphous category known as "low-level" waste. How much of it is produced? What are the various types of waste falling into this category? How long does "low-level" waste remain radioactive? Answers to these questions can be found by looking in more detail at each of the three categories of waste generators—power plants, institutions and industry. We begin with power plants.

High-level waste is irradiated nuclear fuel. Tailings are wastes produced in the process of refining uranium. Transuranic waste is waste contaminated with extremely long-lived nuclides that are heavier than uranium.

LIVING WITHOUT LANDFILLS 9

Nuclear Power Reactors

Nuclear power plants produce energy by splitting, or fissioning, atoms. The fission process gives off a large amount of heat which is used to produce steam and generate electricity.

The main fuel source for nuclear power plants is uranium-235. When first loaded into the reactor, the fuel is only mildly radioactive. It is stacked like poker chips within tubes or rods constructed of zirconium alloy. Clusters of rods, called assemblies, are inserted in water within a steel container, known as the reactor pressure vessel. Figure 1-1 shows the main internal components of a boiling water reactor.

During the fission process, neutrons bombard uranium, which splits into two entirely different and lighter radioactive elements or fission products. In addition, as uranium fissions, it gives off more than two neutrons which, in turn, collide with other uranium nuclei and continue a chain reaction. As a result of the fission process, after several years the fuel rods become quite hot and very radioactive. Depending on the size and type of reactor, 90 tons or more of uranium and highly radioactive fission products sit within the pressure vessel.

In a power reactor, though many waste streams are created, radioactive waste can be generated in only two basic ways: either as a result of the fission process itself, or by neutrons bombarding and altering various materials in the reactor. We shall call the first type "fuel-related waste," and the second type "non-fuel waste."

Fuel-Related Waste

Approximately 240 fission products are created as uranium atoms split. These fission products form within the zirconium tubing, and include cesium, strontium, ruthenium, iodine, cerium, tellurium, technetium, and many others. They are extremely radioactive. Some are solids, while others, such as argon, xenon and krypton, are gases.

Also forming in the fuel rods are lesser quantities of heavy radioactive elements called transuranics. These are created when uranium absorbs neutrons without splitting. Plutonium, americium, curium and neptunium are all transuranics.

If fission products and transuranics remained entirely within fuel rods, no "low-level" waste directly due to the fission process itself would be produced. We would only have high-level waste, that is, spent reactor fuel.

Unfortunately, pin holes and other defects in the zirconium alloy tubing allow certain radionuclides to be washed out of the fuel rods and into the reactor water. The percentage of defective rods is usually small, but enough to cause a "low-level" waste problem. For example, at the Callaway nuclear power plant in Missouri, only 60 rods out of 50,000 within the reactor core are estimated to be defective. Nevertheless, this is estimated to cause about 5,800 curies per year of "low-level" waste to be generated (Union, 1986).

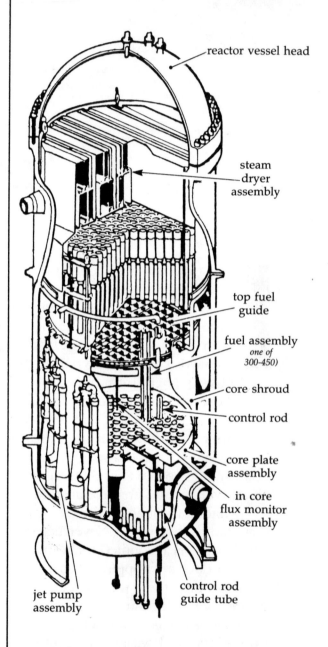

Figure 1-1

Boiling Water Reactor Vessel and Internals

- reactor vessel head
- steam dryer assembly
- top fuel guide
- fuel assembly (one of 300-450)
- core shroud
- control rod
- core plate assembly
- in core flux monitor assembly
- jet pump assembly
- control rod guide tube

Components closest to the nuclear reaction, such as the core shroud, core plate assembly and top fuel guide, develop the greatest amount of long-lived radioactive waste.

These impurities in the water adversely affect the fission process and increase occupational radiation exposures. Reactor water must therefore be cleaned continuously. Leakage from reactor to secondary coolant systems also requires cleanup of the secondary system. Several processes for cleansing the reactor water create "low-level" waste.

Ion exchange resins remove radioactivity by displacing non-radioactive ions initially on the resins with their radioactive counterparts. Reactor water is passed through a cylinder containing the resins, which are plastic, and resemble caviar in consistency. When fully loaded with fission products, the resins are slurried into cylindrical "liners" and dewatered. The liners are transported to waste facilities within upright reusable transport casks.

In some cases, resins are mixed with cement. When cementized resins reabsorb water, the resins swell, crumble and lose compressive strength (Gay,1986). The container will then collapse. Any acid formed when the resins deteriorate will rapidly degrade cement.

Another waste stream, *concentrated liquids*, is formed from evaporating liquid radioactive waste. Concentrated boric acid sludge is a part of this waste stream. Boric acid is added to reactor water to absorb neutrons. As seen in Figure 1-2, in pressurized water reactors, concentrated liquids comprise 34 percent of the "low-level" waste radioactivity, and ion exchange resins account for another 39 percent.

Two types of accidents have occurred with reactor resins and concentrated liquids. At the Three Mile Island reactors in Pennsylvania, fully-loaded resins, solidified with urea formaldehyde and fully exposed to the hot sun, have spontaneously overheated and burst into flames on the storage pad. Cement and vinyl ester-styrene are now used as solidification agents in place of urea formaldehyde. To avoid overheating, new Nuclear Regulatory Commission regulations place an upper cap on the amount of radioactivity that can be contained in an ion exchange resin.

In addition, resins and concentrated liquids generate gases which place tremendous stress on disposal containers. Waste containers designed to last 150 years, and to be leak-proof, apparently contained hydrogen gas too well. These containers, called High Integrity Containers, HIC's, are transported as a liner within a cask. In several cases, gases bulged the airtight containers, making it impossible to remove them from the transport casks without penetrating the containers. In one case, a dry cleaning solvent, and in another, biological activity, caused the HIC containers to bulge when the transport container cover was removed at the Barnwell waste facility.

In another case, at the Arkansas Nuclear One reactor, the ion exchange resin began to spontaneously heat up (to 365°F) when water was removed, causing smoke and steam to be emitted. Because a sewage-like smell was present, Brookhaven scientists conjectured that biological activity was responsible for the heatup (NRC,1986b). To address this problem, new federal regulations allow licensees to build HIC's with vents to allow gas to escape. However, this creates another problem: it is expected that the vents in these containers will be a future avenue for water in-migration and radionuclide leakage.

Radiation from fully-loaded resins has deleterious effects on the resins and containers. Beta and gamma radiation can degrade plastic, weakening the structural integrity (McConnell,1986). Hydrolysis of water can free hydrogen and oxygen gases, and increase the acidity of liquid, forming hydrochloric acid (Jur,1986). Unless the container is vented, increased gases can bulge and weaken the container. Hydrogen, together with beta/gamma radiation, can embrittle and further weaken the high density polyethylene container. When placed under tons of earth in a landfill, the container may fail. If the high integrity containers are constructed of stainless steel, a strong solution of hydrochloric acid can lead to pitting corrosion.

The remaining 27 percent of the fuel-related radioactivity from a reactor is contained in trash (compactible and

Figure 1-2

PWR "Low-Level" Waste
without irradiated components

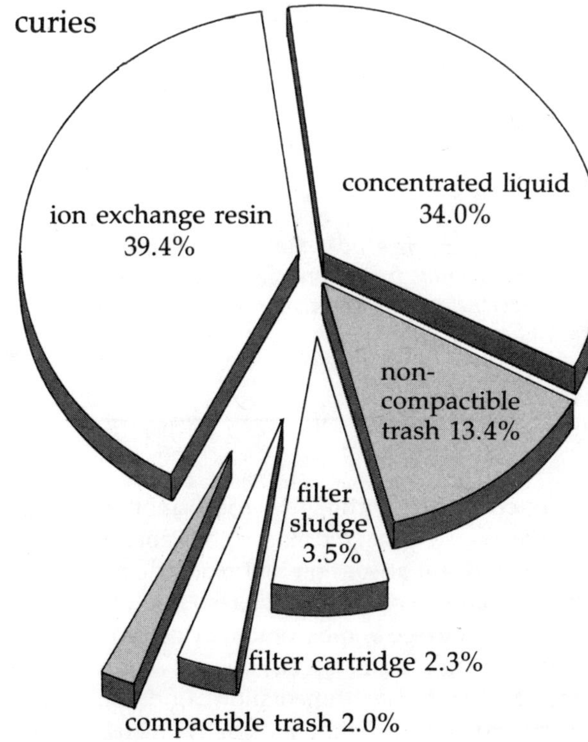

curies

- ion exchange resin 39.4%
- concentrated liquid 34.0%
- non-compactible trash 13.4%
- filter sludge 3.5%
- filter cartridge 2.3%
- compactible trash 2.0%

Excluding irradiated reactor components, concentrated liquids and ion-exchange resins contain over 70% of the "low-level" radioactive waste from nuclear reactors.

non-compactible), filter sludges and cartridge filters.

Trash consists of mops, booties, paper, etc. As seen in Figure 1-2, about 15 percent of the radioactivity is contained in trash, which nevertheless makes up about 60 percent of the "low-level" waste volume. Using supercompactors, trash can be greatly compressed, to a tenth of its former volume, as discussed in Chapter 4. The practice is being pursued by an increasing number of utilities.

Filter sludges and *cartridge filters* also remove radioactivity from reactor water. Cartridge filters, made from cotton or nylon, filter out solid particles. Filter sludge is comprised of powdered resins and cellulose fibers which capture radioactivity and form a thin cake on wire mesh or cloth. The relative radioactivity and volume of these waste streams is shown in Figures 1-2 and 1-3, respectively.

Within the past two years, the Nuclear Regulatory Commission has granted licensees permission to dispose of sludges and contaminated materials with low radioactive concentrations in quite "creative" ways. Contaminated sludges, soil and materials from settling ponds have been land-farmed, that is, spread on earth and plowed into the topsoil. Some contaminated materials have been sent to municipal landfills. Except for the HB Robinson 2 reactor, where approximately 2 curies of settling pond sediment have been transferred to an ash pond on-site, the levels have so far been in the millicurie range (Branagan,1986). As might be expected, however, the number of requests and the amount of radioactivity are expected to grow as disposal costs at radioactive landfills rise.

For how long does fuel-related "low-level" waste remain hazardous? By "hazardous life," we mean the time required for the radioactive concentrations in waste to drop to 100 times maximum permissible concentrations allowed by the Nuclear Regulatory Commission, as specified in its 10 CFR Part 20 regulations. See Table 1-1 and the discussion at the end of this Chapter.

Because of the presence of iodine-129, with a half-life of 17 million years, fuel-related "low-level" wastes, such as ion exchange resins, have a hazardous life of over 100,000 years. Iodine-129 can also be water-soluble. Long-lived radionuclides, such as plutonium and americium isotopes, are also present. For the first 300 to 450 years, the radionuclides strontium-90 and cesium-137, which have approximately 30-year half-lives, present the greatest hazard. The other predominant radionuclides in reactor "low-level" waste are shorter-lived tritium, iron-55, and cobalt-60.

In addition to radioactivity in "low-level" waste, fission products are also released from nuclear reactors in gaseous form. Gases, such as krypton-89 and 90, and xenon-137, can decay to solid radionuclides, strontium-89 and 90, and cesium-137, respectively. Gases can also penetrate porous concrete walls and floors of the reactor building, then breaking down into a range of solid daughter products.

Non-Fuel Waste

Even if all fission products and transuranics were suc-

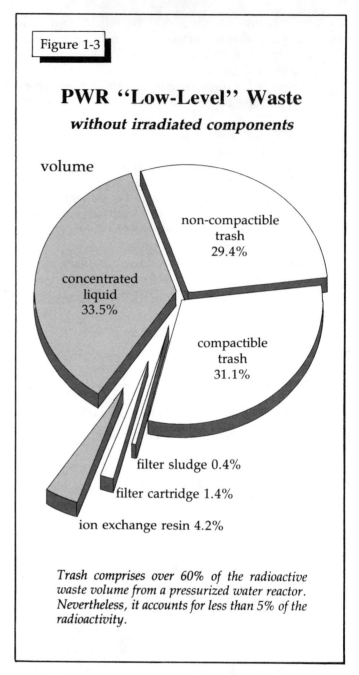

Trash comprises over 60% of the radioactive waste volume from a pressurized water reactor. Nevertheless, it accounts for less than 5% of the radioactivity.

cessfully contained within fuel rods, another source of radioactive waste, in fact the most significant and long-lived source, would still arise: non-fuel reactor components or irradiated components. This major source of "low-level" waste comes from the bombardment by neutrons of reactor components (e.g., fuel channels, control rods, control rod channels and in-core instrumentation). Under such neutron bombardment, normally non-radioactive metals within the reactor become radioactive, or activated. These intensely radioactive components are then removed from the reactor at irregular intervals. The radiation exposure levels from irradiated components can range from 8,000 to 21,000 rads per hour immediately after shutdown (Clymer,1986). To understand what this means, at the high range, an 80-sec-

Table 1-1

Hazardous Life of Representative "Low-Level" Waste Streams
(generated to the year 2020)

waste category	volume (percent)	curies (percent)	hazardous life (years)	major long-lived radionuclide
Ion Exchange Resin (PWR)	1.04	0.1	100,000+	I-129
Concentrated Liquids (PWR)	8.18	0.64	1,000	Ni-63
Non-Fuel Reactor Components (PWR)	1.29	1.42	100,000+	C-14, Ni-59, Cl-36
Reactor Core Shroud (PWR)	0.011[a]	43.7	100,000+	C-14, Ni-59, Cl-36
Reactor Internals (PWR)	0.27	18.4	100,000+	C-14, Ni-59, Cl-36
Decommissioned Ion Exchange Resins (PWR)	0.03	0.5	100,000+	I-129, Pu-239
Compactible Trash (Inst)	5.34	negligible	120	Sr-90
Liquid Scintillation Vials (Inst)	0.12	negligible	15	Co-60
Isotope Production	0.02	0.006	1,000	Sr-90
High Activity Waste	0.1	0.25	100,000+	Ni-59
Tritium Production (New England Nuclear)	0.18	0.41	10,000	C-14
Sealed Sources	negligible	0.1	*	varied

*Although this is only a partial list of the "low-level" waste streams generated to the year 2020 nationwide, the percentages in the table are of the **total** LLW streams.*

Data adapted from (NRC, 1986b)

[a] Can be reduced in volume with supercompactors.

* "Sealed Sources" are small, but intense, single radionuclide sources with hazardous lives from 50 to 250,000 years, depending on the radionuclide contained. Radionuclides include tritium, carbon-14, cobalt-60, plutonium-238, plutonium-239, and americium-241.

ond exposure can lead to death for 50 percent of the exposed population.

The average irradiated component waste volume for a 1,000-megawatt reactor is small, on the order of only 28 cubic meters per year, compared to an average waste volume for all "low-level" waste of 525 cubic meters per year. However, the radioactivity content of this one waste stream is much greater than the sum of all other "low-level" waste streams from the same reactor. As seen in Figure 1-4, the volume of non-fuel reactor components for a pressurized water reactor is just 5 percent, while the fuel-related streams comprise 95 percent. Yet, 85 percent of the total annual low-level radioactivity from a reactor arises from these hot non-fuel components, as seen in Figure 1-5.

Further, the radioactivity in non-fuel reactor components is extremely long-lived. Niobium-94 and nickel-59 have half-lives of 20,000 and 80,000 years, respectively. By comparison, cesium-137 and strontium-90, in other "low-level" waste streams, have half-lives on the order of 30 years, and nickel-63 has a 92-year half-life. In other words, the non-fuel reactor components have hazardous lives on the order of 100,000 years and more, essentially eternity, while the standard "low-level" waste stream has a hazardous life on the order of 300 years.

In addition to irradiated components, neutron bombardment creates other reactor wastes as well. When contaminants in reactor coolant are bombarded, tritium is produced.

Decommissioning Waste

In addition to "low-level" waste produced during each

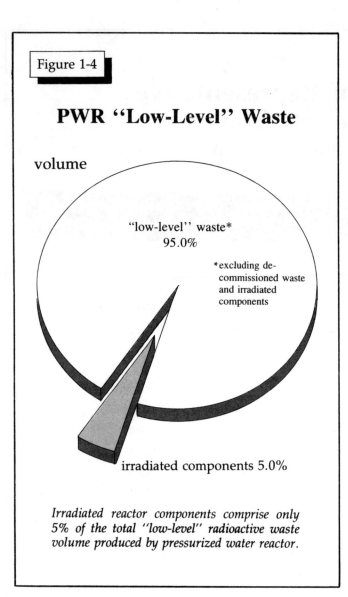

Figure 1-4

PWR "Low-Level" Waste

volume

"low-level" waste*
95.0%

*excluding decommissioned waste and irradiated components

irradiated components 5.0%

Irradiated reactor components comprise only 5% of the total "low-level" radioactive waste volume produced by pressurized water reactor.

As noted above, throughout the course of the reactor's life, neutrons bombard stainless steel within the pressure vessel containing the fuel rods, converting metals like stainless steel alloys into extremely radioactive materials called activation products. The parts of the reactor which become most radioactive lie closest to the center of the nuclear reaction. Neutron activated components contain nickel-59, nickel-63, niobium-94, cobalt-60, and numerous other radionuclides. Some have relatively short half-lives. Cobalt-60 is a strong gamma emitter, but it only has a half-life of 5.27 years. However, as pointed out above, niobium-94 and nickel-59 have half-lives of 20,000 and 80,000 years, respectively.

The reactor internals for a boiling water reactor are shown in Figure 1-1. As can be seen, the lower and upper grid plates, as well as the core shroud, surround the nuclear fuel and nuclear reaction. The lower support columns support the lower grid plate. The lower core barrel surrounds the core shroud, which is in turn surrounded by the thermal shield.

year of a reactor's lifetime, we must also consider the waste resulting from decommissioning a reactor. An average commercial nuclear power plant has an expected operating life of 30 to 40 years, after which it closes down. Theoretically, the facility may then be dismantled following a suitable cooling-off period, probably on the order of 30 years.

The technology for segmenting a large pressurized water reactor vessel has not yet been completely demonstrated. Metal thicknesses of up to 11 inches are "well beyond demonstrated technology for remote cutting (NEA,1981)." The vertical height range (60 feet) also presents a problem. The most promising method for dismantlement, the arc saw, "has not been used in remote applications." Until the technology for dismantlement is developed, it is likely reactors will remain in place.

Surprising as it may seem, the wastes from decommissioning a pressurized water reactor after it shuts down are approximately 100 times more radioactive than the combined total of all the "low-level" waste generated during the 40 years the reactor is in operation.

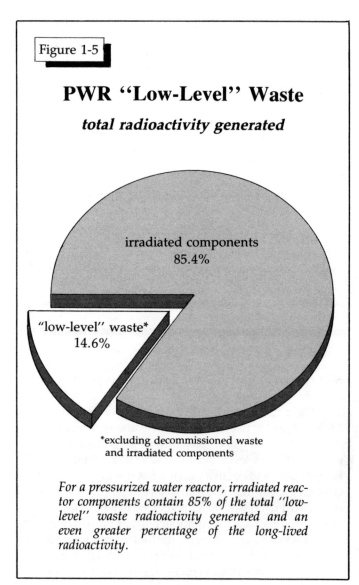

Figure 1-5

PWR "Low-Level" Waste

total radioactivity generated

irradiated components
85.4%

"low-level" waste*
14.6%

*excluding decommissioned waste and irradiated components

For a pressurized water reactor, irradiated reactor components contain 85% of the total "low-level" waste radioactivity generated and an even greater percentage of the long-lived radioactivity.

What is remarkable about these internals is the contrast between how small a volume they take up, and how large an amount of radioactivity they contain. All these components, with a volume of 150 cubic meters, comprise only one-eighth of the total decommissioning waste volume. Expressed in another way, the reactor internals have a total volume that is equivalent to just one-quarter of the "low-level" waste produced by a pressurized water reactor in a single year. Nevertheless, these same 150 cubic meters of reactor internals **constitute 99.5 percent of the total radioactivity in a decommissioned pressurized water reactor.** The total radioactivity in activated metals in a reactor which has operated for 40 years, 4.6 million curies, far exceeds the radioactivity in "low-level" radioactive waste produced annually by a pressurized water reactor, less than 2,000 curies a year. The same arguments hold true for a boiling water reactor.

In addition to activated metals, other forms of "low-level" waste—ion exchange resins, evaporator bottoms and activated concrete—arise when a reactor is decommissioned. In order to reduce occupational radiation exposures, pipes and metal surfaces are rinsed with chemicals called chelating agents to partially remove surface contamination. Chelating agents remove radionuclides like cobalt-60 from surfaces by forming molecule complexes with the radionuclides. The complexes are large, making it difficult for them to "cling" to anything else. This means the complexes of chelating agents and radionuclides can be easily washed away. The chemical mix is passed through ion exchange resins where the radionuclides are removed. The ion exchange resins, with heavy concentrations of chelating agents, then become "low-level" waste.

The presence of chelating agents in landfills has become a major concern in recent years. These chemicals

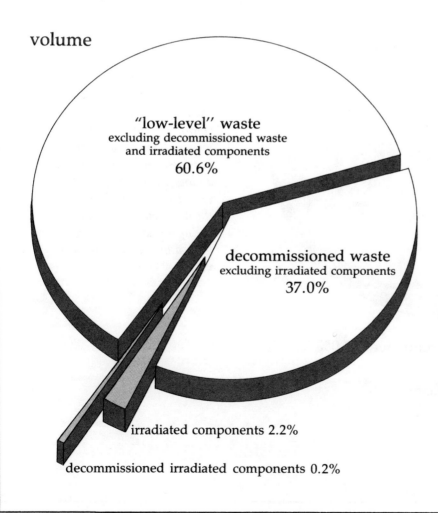

Figure 1-6

Total BWR Waste

volume

"low-level" waste excluding decommissioned waste and irradiated components 60.6%

decommissioned waste excluding irradiated components 37.0%

irradiated components 2.2%

decommissioned irradiated components 0.2%

When "low-level" waste from decommissioned reactors is taken into account, irradiated reactor components constitute a mere 2.4% of the "low-level" waste volume.

have combined with other radionuclides in the landfills, permitting the migration of radioactive materials out of burial trenches and off-site. Concentrations of chelating agents on the order of only parts per million have allowed previously "fixed" cobalt-60 and strontium-90 to move from trenches (Dayal,1983). For more information, see page 35.

In Figure 1-6, the volumes of four boiling water reactor waste streams are compared. We assume here that a 1000-megawatt reactor operates for 40 years and is then decommissioned. As is seen, if we add all non-fuel reactor components from 40 years of operation, and all the reactor internals from decommissioning, the total comprises a mere 2.4 percent of the waste volume. All other "low-level" waste from 40 years of operation and from decommissioning makes up 97.6 percent of the volume.

In comparing the radioactivity (see Figure 1-7), the picture is reversed. Non-fuel reactor components and reactor internals account for 98.6 percent of the radioactivity, made up primarily of long-lived radionuclides. Thus, a small volume of radioactive materials accounts for a very large portion of the radioactivity from nuclear power reactors.

One way to graphically see the long-lived nature of reactor internals compared to other components of "low-level" waste is to track pressurized water reactor waste streams as a function of time. This is done in Figure 1-8. Non-fuel reactor components and reactor internals are com-

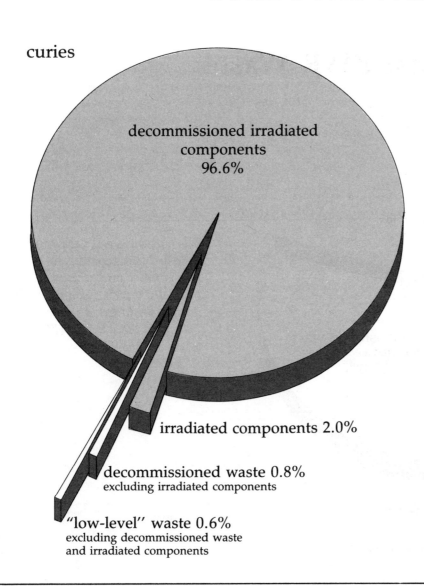

Figure 1-7

Total BWR Waste

curies

decommissioned irradiated components 96.6%

irradiated components 2.0%

decommissioned waste 0.8%
excluding irradiated components

"low-level" waste 0.6%
excluding decommissioned waste and irradiated components

While irradiated components constitute a mere 2.4% of the "low-level" waste volume, they account for 98.6% of the total radioactivity. Excluding these components from a "low-level" waste facility would dramatically reduce the maintenance and monitoring period from tens of thousands to hundreds of years.

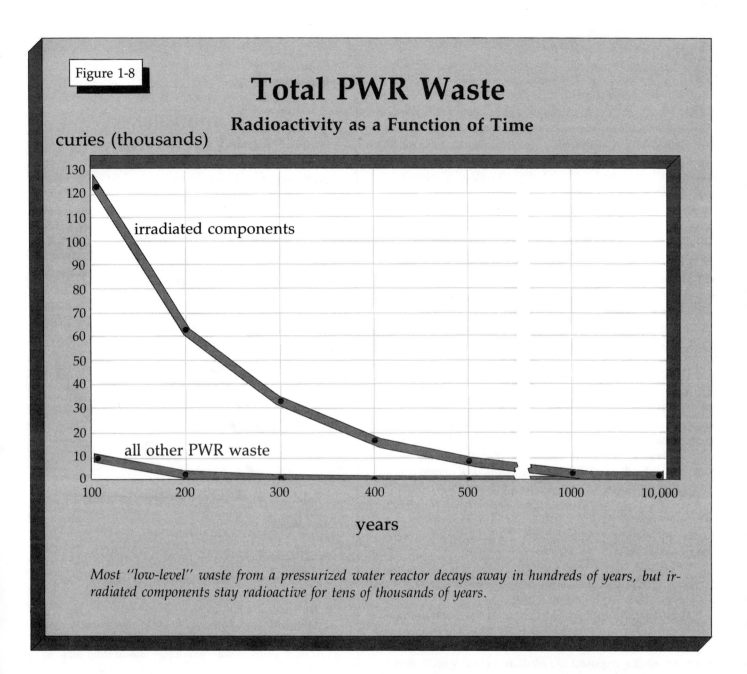

Most "low-level" waste from a pressurized water reactor decays away in hundreds of years, but irradiated components stay radioactive for tens of thousands of years.

bined into one category, irradiated components, and the time scale ranges from 100 years to 10,000 years.

Even at 100 years, the radioactivity due to internal components is at least a factor of 10 greater than the total fuel-related "low-level" waste produced by the same reactor. Because of the presence of cesium-137 and strontium-90, all reactor waste streams remain radioactive and hazardous for at least 300 years. But, as is seen in the graph, even after 10,000 years irradiated components remain radioactive and hazardous due to the presence of nickel-59, nickel-63 and niobium-94. One way to gain a perspective on what 10,000 years means is by stretching such a span backward in time. Ten thousand years ago was the period of Cro-Magnon man. It is highly conjectural to predict what will occur 100 years into the future, let alone 10,000.

Reactor Waste—A Summary

Before going on to discuss institutions, the second of our three categories of waste generators, let us briefly recapitulate what we have discovered about low-level wastes from power plants:

• An average reactor produces both fuel-related and non-fuel waste during each year of its 40-year lifetime.

• Fuel-related waste results from cleansing fission products out of reactor water, whereas non-fuel waste is produced when neutrons from the fission process bombard normally non-radioactive metals and activate them.

• In the fuel-related category, concentrated liquids and ion exchange resins account for most of the radioactivity, while trash accounts for most of the volume. Overall, the fuel-related category is high in volume but proportionately lower in radioactivity.

• The non-fuel category, on the other hand, is low in volume and quite high in radioactivity. Non-fuel radioactivity is found in activated reactor components and is long-lived.

• The waste from decommissioning a nuclear plant is vastly more radioactive than all other low-level waste produced during the reactor's 40-year lifetime. Most of this decommissioning radioactivity is concentrated in a small volume of reactor internals. This waste is very long-lived.

Institutional Waste

Universities, medical schools, research laboratories and hospitals are broadly classed as institutional waste generators. They constitute almost the entirety of the 20,000 waste generators in the United States. Each year, about 100 million nuclear medical procedures are performed in this country, and the rate is increasing by 10 percent a year (ACP,1984). Further, about 30 percent of all biomedical and cancer research uses radioactive markers.

Nevertheless, institutional generators only produce half as much radioactivity as the waste from *one* 1,000-megawatt nuclear power reactor, even excluding non-fuel reactor components. In general, two radionuclides in institutional waste are significant: tritium and carbon-14, the latter long-lived radionuclide being used primarily in universities and research institutions. Both radionuclides are being replaced with phosphorus-32, used as a marker of DNA. Since phosphorus-32 has a half-life of only 14.3 days, the waste problem due to institutions is being decreased to insignificant levels.

Institutional waste can be divided into four broad subcategories: liquid scintillation vials, animal carcasses (biowaste), absorbed liquids and trash.

Liquid scintillation devices are used to count atomic particles, that is, to measure radioactivity. They are based on the fact that flashes of light (scintillations) occur when certain materials are exposed to radiation. Liquid scintillation counting techniques used in biological research employ radionuclides such as tritium, carbon-14 and cobalt-60 mixed with an organic solvent carrier like toluene, benzene or xylene to detect radioactivity in blood or urine. These solvents are classed as toxic chemicals, and have been known to burst into flames during transport to waste facilities. They are presently being incinerated by Quadrex Corporation in Gainesville, Florida.

Medical researchers have lately been converting to carriers that can be mixed with water (water-miscible) in scintillation vials. However, even these counters still employ radionuclides. As shown in Figure 1-9, liquid scintillation vials make up a mere 0.2 percent of institutional radioactivity, which, in turn, makes up only a fraction of a percent of the radioactive waste produced annually in the United States.

Radioactive animal carcasses, or biowaste, are pro-

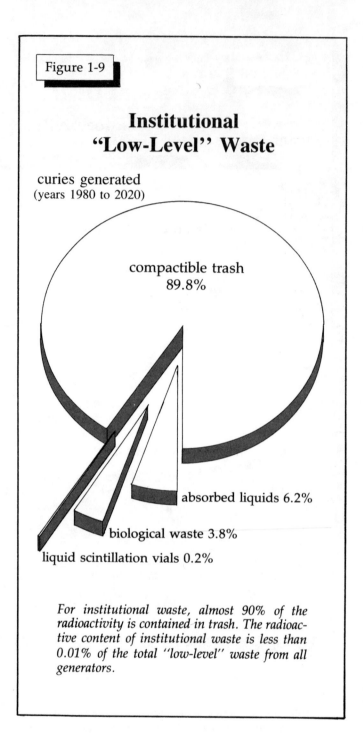

Figure 1-9

Institutional "Low-Level" Waste

curies generated
(years 1980 to 2020)

compactible trash 89.8%

absorbed liquids 6.2%

biological waste 3.8%

liquid scintillation vials 0.2%

For institutional waste, almost 90% of the radioactivity is contained in trash. The radioactive content of institutional waste is less than 0.01% of the total "low-level" waste from all generators.

duced by research programs at universities. Chemical compounds that are being considered for use as a human or veterinary drug are tagged with tritium or carbon-14 and injected into research animals in order to study how the chemical compounds behave. Afterwards, the research animals, containing trace quantities of radioactivity, are killed. They are then categorized as "waste" and are either incinerated or packed in lime and buried at radioactive landfills.

Absorbed liquids contain a broader range of radionuclides in aqueous and organic solvents than the above types of institutional radioactive waste.

By far the largest volume and radioactivity in the institutional category is in *compactible trash*. This consists of paper, gloves, syringes and labware. Compactible trash makes up 90 percent of the radioactivity in institutional waste. Though this trash is compressed at laboratories, it could be even further compressed, to 10 percent of the original volume. This could be done by supercompactors, which are discussed in Chapter 4. The relative radioactivity and volumes of institutional waste streams are shown in Figures 1-9 and 1-10.

Because of the expense of disposing of a large trash volume in a radioactive landfill, institutional waste generators have promoted the burning of waste at large centralized incinerators. US Ecology, the company that operates the Richland, Washington waste landfill, and Babcock & Wilcox, the designer of the now-famous Three Mile Island nuclear reactor, have proposed radioactive incinerators in Bladen County, North Carolina and Parks Township, Pennsylvania, respectively. The license application for the proposed North Carolina incinerator has now been rejected by state agencies.

The principal radionuclides in the waste to be burned, tritium and carbon-14, would become radioactive water vapor and carbon dioxide, respectively, and would enter the environment in small quantities. Local residents obviously oppose this practice of dispersing radioactivity into the air. Another serious problem has arisen with incineration: the possible production of an extremely toxic chemical, dioxin, when polyvinyl chloride plastic and paper are burned (RWC,1986). This is a potential problem for all, not just radioactive, incinerators.

Since the adoption of new federal regulations in December of 1981 (see page 45), the overall volume and radioactivity of institutional waste has declined dramatically, even at a time when more radionuclides are being used by medical and research institutions. The reason is quite simple: the regulations permit institutions to discharge waste containing concentrations of tritium and carbon-14 lower than 0.05 microcuries per gram down the drain or in municipal landfills. The scintillation volumes disposed of at waste facilities show a very clear and unmistakable trend: 199,000 cubic feet in 1980; 101,000 cubic feet in 1982; 80,000 cubic feet in 1983; and just 40,000 cubic feet in 1984. Absorbed liquids have also shown a marked decline.

Rather than being absorbed and disposed of, radioactive liquids such as tritium, carbon-14 and others are being poured down the drain. As a result, more radioactivity is showing up in municipal sludge near manufacturers, hospitals and research institutions, and is generally increasing background radiation levels. Radioisotopes which have been detected include iodine-131, cesium-137 and gadolinium-153. No systematic monitoring or enforcement has been carried out by the U.S. Nuclear Regulatory Commission, though a memorandum was sent to all licensees informing them of three incidences of high levels of radioactivity in sewage sludge (NRC,1984a).

Industrial Waste

The products and services making use of radioactivity in this country are quite varied. They range from exotic items like uranium bullets to commonplace items such as smoke detectors, and include radiopharmaceuticals, radiography sources, static eliminators, well-logging sources and glow-in-the-dark EXIT signs. Some manufacturers produce radioisotopes, but most use and repackage radioisotopes which are produced by the federal government. In contrast with the medical and research institutions category, which has a large number of generators, the number of industrial licensees is not large. Thus, where possible, we identify the names and locations of major manufacturers.

Industrial generators can be divided into several subca-

Figure 1-10

Institutional "Low-Level" Waste

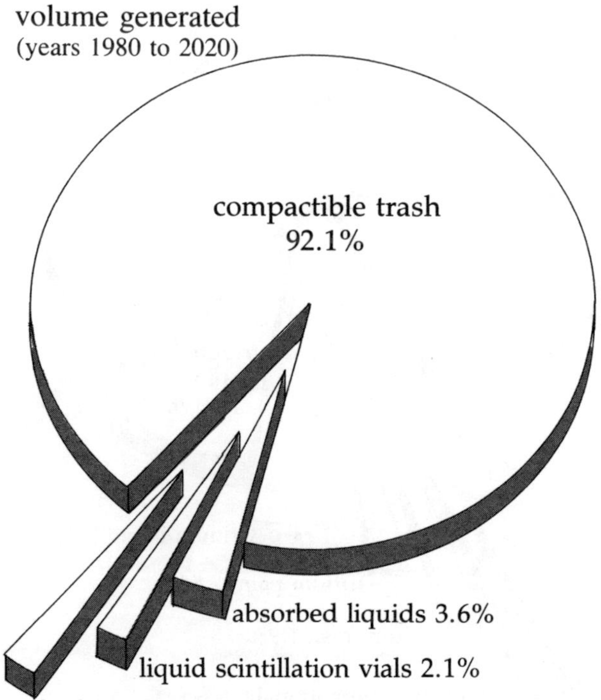

volume generated
(years 1980 to 2020)

- compactible trash 92.1%
- absorbed liquids 3.6%
- liquid scintillation vials 2.1%
- biological waste 2.1%

Almost the entire volume of "low-level" waste from research institutions and hospitals consists of contaminated paper and plastic trash. Compactible trash can be compressed into a much smaller volume and stored.

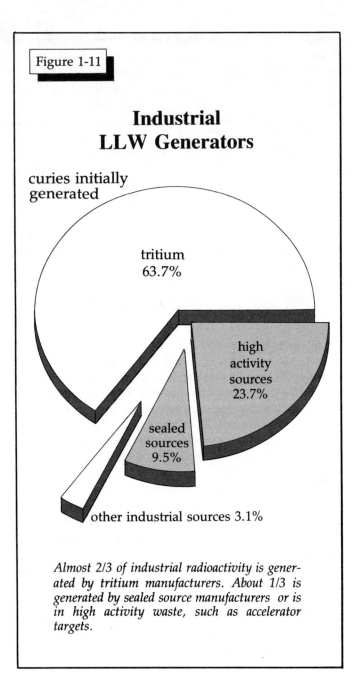

Figure 1-11

Industrial LLW Generators

curies initially generated

- tritium 63.7%
- high activity sources 23.7%
- sealed sources 9.5%
- other industrial sources 3.1%

Almost 2/3 of industrial radioactivity is generated by tritium manufacturers. About 1/3 is generated by sealed source manufacturers or is in high activity waste, such as accelerator targets.

sealed in glass bulbs for disposal.

Since tritium has a half-life of 12.3 years, it is feasible to store tritiated waste until it decays to non-hazardous levels, on the order of 120 years. Unfortunately, however, in the case of New England Nuclear, carbon-14 and tritium wastes are commingled, or mixed, turning a 120-year problem into a 57,300-year problem because of the long half-life of carbon-14. As seen in Figures 1-12 and 1-13, a Midwest tritium manufacturer, Amersham, is a much smaller generator of tritium and carbon-14 waste.

The other tritium waste generators producing specific tritium products are broken into the following subcategories:

• tritium in paint and plating (illuminated signs)

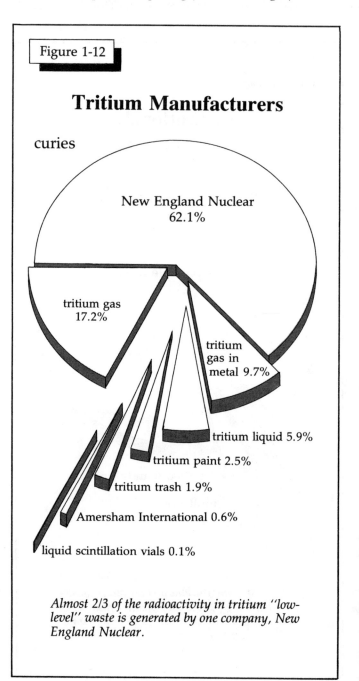

Figure 1-12

Tritium Manufacturers

curies

- New England Nuclear 62.1%
- tritium gas 17.2%
- tritium gas in metal 9.7%
- tritium liquid 5.9%
- tritium paint 2.5%
- tritium trash 1.9%
- Amersham International 0.6%
- liquid scintillation vials 0.1%

Almost 2/3 of the radioactivity in tritium "low-level" waste is generated by one company, New England Nuclear.

tegories: tritium manufacturers, high activity generators, sealed source manufacturers and other industrial generators.

As a group, tritium manufacturers produce two-thirds of the radioactivity initially generated in industrial waste, as seen in Figure 1-11. One company dominates the production of tritium and carbon-14 for university and medical research, New England Nuclear based in Billerica, Massachusetts. Tritium, a by-product of nuclear weapons production, is shipped to New England Nuclear, where it is repackaged in various physical and chemical forms. As seen in Figures 1-12 and 1-13, respectively, New England Nuclear accounts for two-thirds of the radioactivity and three-quarters of the volume shipped to waste facilities by tritium manufacturers. Most of this radioactivity is in a gaseous form

- tritium in a gas
- absorbed aqueous liquid
- tritium in trash
- liquid scintillation vials
- tritium gas absorbed in metal

The relative radioactivity and volume produced by each of these tritium waste generators is shown in Figures 1-12 and 1-13.

Though tritium manufacturers are responsible for two-thirds of the industrial radioactivity initially generated, because of tritium's short half-life (compared to other industrial radionuclides), this waste accounts for only 18 percent of the radioactivity remaining in industrial waste after 100 years. This can be seen by comparing Figures 1-11 and 1-14.

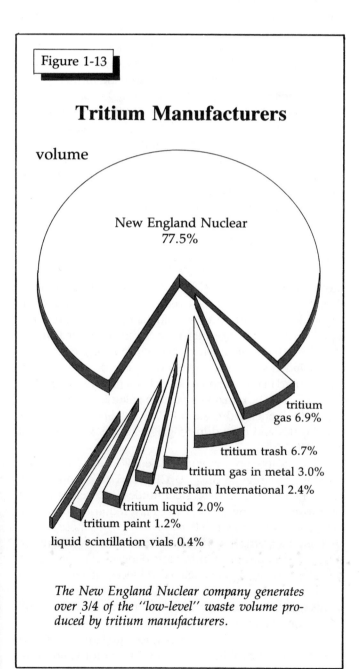

The New England Nuclear company generates over 3/4 of the "low-level" waste volume produced by tritium manufacturers.

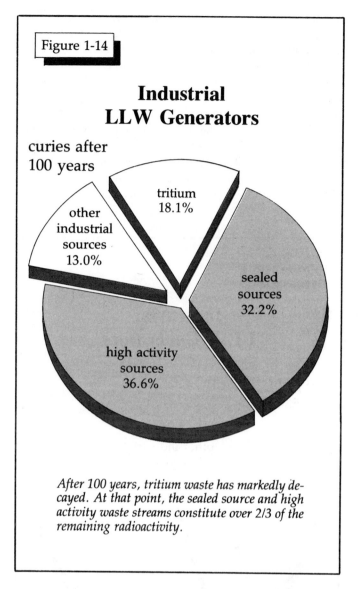

After 100 years, tritium waste has markedly decayed. At that point, the sealed source and high activity waste streams constitute over 2/3 of the remaining radioactivity.

The next largest source of industrial radioactive waste is high activity generators, labelled high activity. This subcategory includes activated metals produced by accelerators, research reactors, and neutron generators. High activity contains essentially the same long-lived radionuclides as are found in reactor internals, though the volume is small, on the order of 75 cubic meters a year. The radioactive concentrations, however, are very high.

As seen in Figure 1-11, the high activity waste stream accounts for almost one-quarter of industrial "low-level" waste radioactivity when initially generated, but one-third of the radioactivity present 100 years later (see Figure 1-14). As shown in Figure 1-15, the volume of high activity waste compared to other industrial waste is miniscule, about 0.4 percent of the industrial "low-level" waste volume.

The third largest group of industrial waste generators is lumped together as sealed source generators. Sealed sources have a large variety of applications: neutron generators, medical and industrial irradiators, well-logging and radiography sources, and static eliminators. The widest use is in smoke detectors. The radionuclides employed in sealed sources are tritium, carbon-14, cobalt-60, nickel-63, strontium-90, cesium-137, plutonium-238 and 239, and americium-241. The volume of individual sealed sources is extremely small, about the size of an index finger, but the average radioactive content is large, about 10 curies. Major manufacturers include NRD Corp in the Northeast, Minnesota, Mining and Manufacturing in Minnesota, and the General Electric Vallecitos Nuclear Center in California.

It is worth noting that sealed source manufacturers obtain their radioisotopes from the federal government. For example, Minnesota, Mining and Manufacturing obtains cesium-137 from Oak Ridge National Laboratory, which, in turn, receives this radioisotope from high-level waste generated by nuclear weapons production. Americium-241 is also obtained from nuclear weapons production. Plutonium-241 in nuclear weapons decays to americium-241 which can then be separated from other plutonium isotopes. Americium-241 is extracted from recycled nuclear warheads at the Department of Energy's Rocky Flats Plant. The production of americium-241 sources for smoke detectors north of Buffalo has led to contamination of sewage sludge in the Tonawanda and Grand Island municipal sewer systems (DOH,1986).

As seen in Figures 1-11 and 1-14, sealed sources represent 9.5 percent of the industrial radioactivity initially generated, but about one-third of the radioactivity still present after 100 years. The volume is insignificant.

Another major source of industrial "low-level" waste is Cintichem, a subsidiary of Hoffman-LaRoche, in Tuxedo, New York. Cintichem, which took over the former Union Carbide facility, produces molybdenum-99 for hospitals. Molybdenum-99 decays to technetium-99m, which is used extensively as a radioisotope tracer in patient diagnosis.

The production of molybdenum-99 is essentially identical to the extraction of plutonium and uranium from irradiated fuel at government reprocessing facilities. At Cintichem, bottles lined with uranium-235 are irradiated in a reactor. Molybdenum-99 is chemically separated from the fissioned uranium.

Under the Nuclear Regulatory Commission's high-level waste regulations, 10 CFR Part 60, the liquid waste containing fission products generated in this manner is high-level waste. However, the NRC has chosen to view it as "low-level" waste because of the expense and the importance of the product in health care (Cunningham,1985). A part of the liquid waste is transported to the Savannah River Plant in Aiken, South Carolina, where fission products are removed and included in defense high-level waste. Recovered uranium is used in the Savannah River Plant plutonium production reactors. The remaining portion of liquid waste, including all the generated cesium-137, is solidified in stainless steel bottles and placed in 55-gallon drums with assorted trash (NRC,1983).

This points up a major deficiency in the NRC waste management regulations, 10 CFR Part 61. Cintichem's highly radioactive stainless steel bottles themselves would fall into an NRC "low-level" waste class requiring more care than many other "low-level" wastes. But when these bottles are mixed with trash in 55-gallon drums, the average radioactive content just fits within a less rigorous class. Since the NRC has no regulations against dilution, generators can move from more to less restictive management requirements simply by diluting waste with enough trash. For a full discussion of NRC regulations, see Chapter 3.

Several industrial waste generators are lumped into the "other" subcategory in Figure 1-15. Except for Cintichem, these companies are generators of high volume, low radio-

Figure 1-15

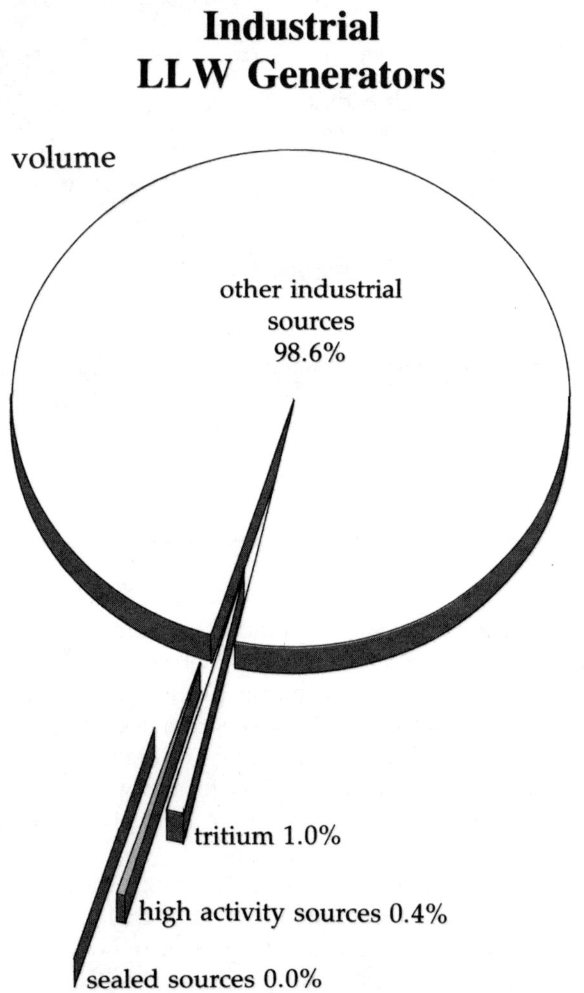

Industrial LLW Generators

volume

- other industrial sources 98.6%
- tritium 1.0%
- high activity sources 0.4%
- sealed sources 0.0%

Even though sealed source and high activity generators constitute over 2/3 of the radioactivity remaining after 100 years, they comprise a mere 0.4% of the waste volume.

active concentration waste. These waste streams represent industrial research waste from corporations like Dow Chemical, pharmaceutical companies such as Squibb and Abbott, and testing and analytical laboratories. These waste streams are quite similar to university and medical research waste. Also included are two waste streams arising from the fabrication and production of uranium, such as for uranium armor-piercing shells. The largest volume producer, Nuclear Metals, Inc., is located in Massachusetts.

Finally, waste from nuclear submarines is also included in the "other" category. Similar to commercial power reactors, nuclear submarines produce radioactive trash and "wet" waste, such as ion exchange resins. The total nuclear generating capacity from all submarine reactors is approximately equal to one large commercial power reactor, but the production of waste is at a much lower rate, perhaps because naval fuel is manufactured to higher standards than commercial fuel and does not develop the same percentage of pin holes.

Naval reactor "low-level" waste is unloaded at ports in Washington, California, South Carolina, Virginia, Connecticut and New Hampshire/Maine. It is not clear whether commercial waste facilities must legally accept this waste, or whether it will go to Department of Energy landfills in Tennessee, South Carolina, Nevada or Washington. Submarine waste has been going to commercial landfills at Barnwell, South Carolina and Richland, Washington.

As seen in Figures 1-11 and 1-15, the "other" category generates 98.6 percent of the industrial waste volume, but a mere 3.1 percent of the radioactivity.

Purposefully excluded from consideration here is waste generated during cleanup operations at West Valley,

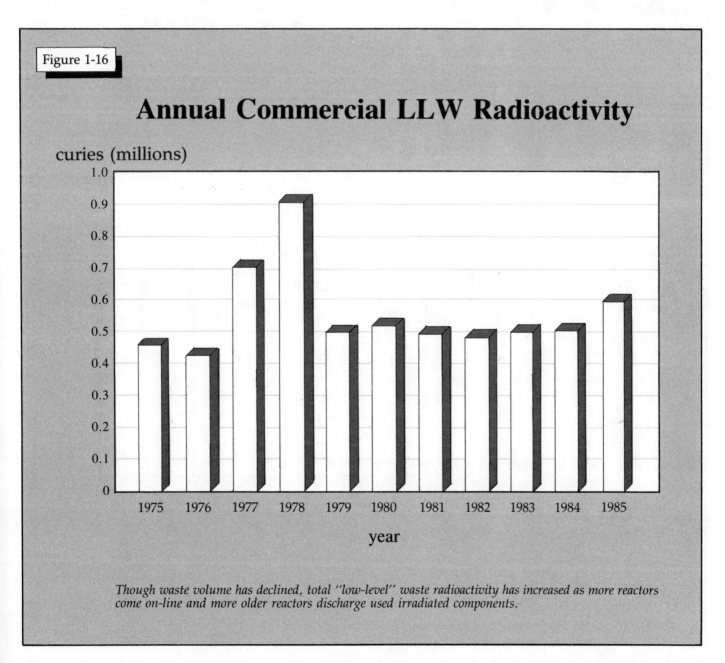

Figure 1-16

Annual Commercial LLW Radioactivity

Though waste volume has declined, total "low-level" waste radioactivity has increased as more reactors come on-line and more older reactors discharge used irradiated components.

New York, and any future reprocessing waste. At West Valley, the Department of Energy is solidifying liquid high-level waste from a now-closed commercial reprocessing operation formerly run by a subsidiary of Getty Oil. This process will generate about 60,000 curies of "low-level" and transuranic-contaminated waste, which will either go to a regional or New York State waste facility, or remain at West Valley. This "low-level" waste will largely be in the form of concrete in drums which the Department of Energy would like to place in an above ground pyramid, or tumulus, at the site (see Chapter 4).

Also excluded from consideration here is waste from future reprocessing operations. The NRC does include this source (NRC,1986a), but we believe it to be too speculative to warrant inclusion in this report. Also not considered here, though it is more likely than future reprocessing, is the waste produced if the entire nuclear industry closed down immediately, and the major production of "low-level" waste ceased.

Hazardous Life

In developing new regulations for "low-level" waste, 10 CFR Part 61 (see Chapter 3), the Nuclear Regulatory Commission divided "low-level" waste into three categories, classes A, B and C. Waste that was high-level waste, uranium mill tailings, or weapons production waste, was not covered by Part 61. Waste that was classified as greater than

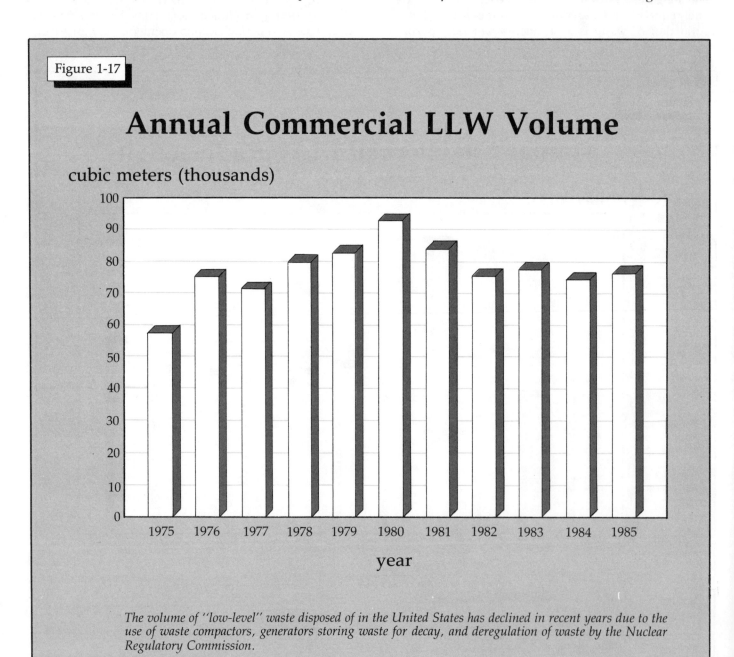

Figure 1-17

Annual Commercial LLW Volume

The volume of "low-level" waste disposed of in the United States has declined in recent years due to the use of waste compactors, generators storing waste for decay, and deregulation of waste by the Nuclear Regulatory Commission.

class C could not be disposed of in a near-surface disposal facility, though the NRC has initiated proceedings which may incorporate more high-level waste into the "low-level" waste category (see Chapter 3).

Classes A, B and C were based on computer models of landfill leakage that included many general assumptions about the characteristics of waste and waste sites, and about how future generations would behave upon discovering waste. Putting all the assumptions in the pot, and applying different regulatory requirements to the different waste classes, the NRC calculated that the dose due to leakage from a waste facility would never exceed 25 millirems per year to an individual at the boundary, and would not exceed a total radiation dose of 500 millirems to an inadvertent intruder. This assumed the waste facility would be guarded for 100 years after closure.

Completely bypassed in this new classification scheme was a section of the NRC regulations, 10 CFR Part 20, which had served for over 20 years as the yardstick for hazard. Part 20 defines the maximum permissible concentration (MPC) for any radionuclide, or combination of radionuclides, in air or water that can reach the public. MPC's are calculated so that, theoretically, no member of the public receives a whole body radiation dose greater than 500 millirems per year, or a dose to any organ greater than 1,500 millirems per year. Therefore MPC is a measure of the hazard of radioactive waste which takes into account biological uptake and retention, and the effect of radiation on humans.

Another, more imprecise, indicator is the "ten times the half-life" rule. In a period of time ten times the half-life, the radioactivity of a radionuclide drops by a factor of 1,000. This is a less precise measure because it does not factor in the concentration of individual radionuclides or the radiation hazard.

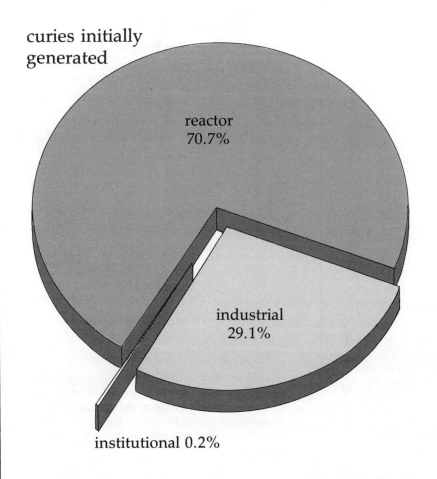

Figure 1-18

"Low-Level" Waste
without decommissioned waste

curies initially generated

reactor 70.7%

industrial 29.1%

institutional 0.2%

Projecting waste generation to the year 2020 for the country as a whole, nuclear power reactor waste will constitute about 70% of the radioactivity, and industry 30%. Institutional waste will contain about 0.2% of the radioactivity. Long-lived waste from decommissioned reactors is not included in these percentages.

If one defines the hazardous life of radioactive waste by how long it takes for its concentration to drop to MPC, one has an approximate indicator of the time a particular waste stream must be monitored. Using MPC as a measure, the hazardous lives of different "low-level" waste streams are listed in Table 1-1. As can be seen, these vary from 15 years to over 100,000 years. Ion exchange resins, used to clean reactor water, have a hazardous life of over 100,000 years, due to the presence of iodine-129. Irradiated reactor components and resins from the decommissioning of reactors have hazardous lives in excess of 100,000 years due to the presence of chlorine-36, nickel-59 and iodine-129. Clearly ion exchange resins and irradiated components deserve to be treated as high-level waste. In Table 1-1, some of the more radioactive waste streams which warrant particular care are separated out. Also listed for the various waste streams are the percentages they account for out of total U.S. waste volume and radioactivity to the year 2020, along with the major long-lived radionuclides.

The Big Picture:
Estimating Total "Low-Level" Waste

Now that we have examined each of the three categories of generators in detail, it is useful to put the pieces together to gain an overall picture of "low-level" waste generation trends in the United States.

The annual radioactivity and volume of commercial "low-level" waste generated since 1980 may be found in Figures 1-16 and 1-17, respectively. Included are the more than 20,000 companies, institutions, laboratories and government facilities that use radioactive materials and generate low-level waste. The waste ranges from mildly radioactive trash to highly radioactive stainless steel from the interior of a nuclear reactor, from microcuries per cubic meter to thousands of curies per cubic meter. As noted above, most of the radioactivity is contained in a small volume that is part of the waste generated by fewer than 100 licencees, most of them utilities.

In 1985, a total "low-level" waste volume of approximately 80,000 cubic meters was generated in the U.S., down from a peak of 92,400 cubic meters in 1980. This decline in volume is due to compaction of waste and, as described above, the increasing disposal of unregulated waste in landfills and down drains beginning in 1981. Despite this decline in volume, the total radioactivity disposed of has increased since 1982, reaching 600,000 curies in 1985. Unless other wastes are ruled below regulatory concern, this upward trend is expected to continue as more nuclear reactors operate and more reactor internals become hotter over time.

And how much waste will have to be managed over the next three or four decades? Between the years 1980 and 2020, for the nation as a whole, about one-half BILLION curies of "low-level" waste will have been generated by institutional and industrial generators and power reactors, excluding shorter-lived radionuclides and including waste from decommissioning nuclear reactors. When leakage of one billion*th* of a curie is cause for great alarm in a laboratory or hospital, one billion curies is almost too great to comprehend.

How does one arrive at the figure "one-half BILLION curies?" The calculation is based on several assumptions. First, a judgment must be made about how many power reactors will operate between now and the year 2020. Except for certain obvious cancellations, we assume the capacity on-line or under construction in 1984 is the total capacity between now and the year 2020. We assume therefore an electrical capacity of approximately 70,000 megawatts from pressurized water reactors and 32,000 megawatts from boiling water reactors. Perhaps some power

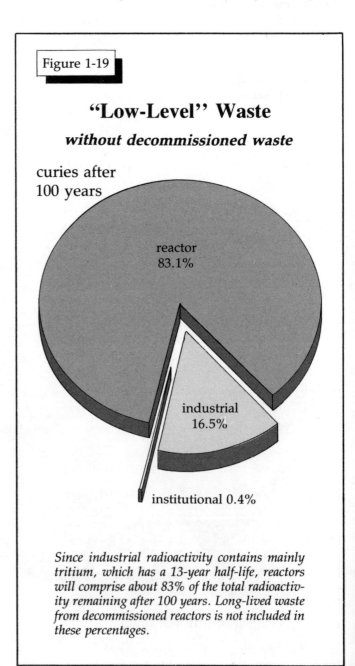

Figure 1-19

"Low-Level" Waste
without decommissioned waste

curies after 100 years

reactor 83.1%

industrial 16.5%

institutional 0.4%

Since industrial radioactivity contains mainly tritium, which has a 13-year half-life, reactors will comprise about 83% of the total radioactivity remaining after 100 years. Long-lived waste from decommissioned reactors is not included in these percentages.

reactors will operate only 30 rather than 40 years, or perhaps some new reactors will go on line in the year 2005. These perturbations are impossible to project and are ignored. We assume the above fixed capacity; shut down reactors are replaced with new capacity under this assumption.

Another assumption relates to how much waste each reactor produces. In general, smaller reactors will produce more activated metal than larger reactors per unit of capacity because neutrons are more likely to strike the core shroud rather than the fuel in a smaller reactor. We ignore this subtlety as well and assume that all pressurized and boiling water reactors have a 1,000-megawatt capacity. This assumption, then, underestimates the radioactivity in decommissioning waste.

Growth in institutional and industrial waste generation to the year 2,000, and constant thereafter, is factored into the calculation, as projected by NRC's *Update* (NRC, 1985). Finally, we consider "low-level" waste for the nation as a whole, and do not particularize waste generation to a specific state or region.

For the nation as a whole, and excluding decommissioning waste for the moment, the radioactivity initially generated by sector is shown in Figure 1-18. As seen, power reactors produce about 70 percent of "low-level" waste radioactivity, and industrial generators about 30 percent. Institutional generators produce a mere 0.2 percent of the radioactivity. At 100 years after waste generation (see Figure 1-19), reactor waste accounts for 83 percent of total radioactivity and industrial waste about 16.5 percent of the remain-

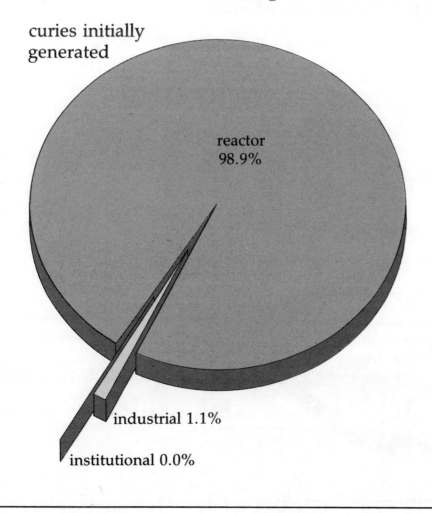

Figure 1-20

"Low-Level" Waste
including decommissioned waste

curies initially generated

reactor 98.9%

industrial 1.1%

institutional 0.0%

When waste from decommissioned reactors is included in projections of total waste generation through 2020, nuclear power plants will generate almost 99% of the total commercial radioactivity. Institutions will generate a mere 0.008%.

LIVING WITHOUT LANDFILLS 27

ing radioactivity, reflecting the fact that non-fuel reactor components are very long-lived and that tritium in industrial waste has a half-life of 12.3 years.

Including decommissioning waste greatly alters the equation. For this analysis, it is not necessary to further assume that reactors are dismantled when decommissioned, only that the waste is generated. With decommissioning waste included, power reactors produce almost 99 percent of "low-level" waste radioactivity, industry about 1 percent and institutional waste (with about 0.008 percent) gets lost in the noise (see Figure 1-20).

In terms of waste volume, and excluding decommissioning waste, power reactors produce 55 percent of the total volume, industry 36 percent and institutions 9 percent, as shown in Figure 1-21. Including decommissioning waste, the volume produced by power reactors increases to 71 percent, and industry and institutions together produce 29 percent of the "low-level" waste volume.

Total radioactivity as a function of time, from 100 years after waste generation to 10,000 years after waste generation, is plotted in Figure 1-22. Institutional and industrial waste are lumped into one category, though institutional waste is the much smaller component, comprising less than 0.008 percent of the total "low-level" waste radioactivity produced in the country. The initial time, 100 years, is the end of the institutional control period, after which the site is likely to be unregulated and uncontrolled. As is seen, even 10,000 years after waste generation, 200,000 curies from 115 power reactors, almost entirely activated components, still remain. Who will monitor and maintain these wastes?

These results provoke the obvious question: If 99 percent of "low-level" waste radioactivity is generated at 115 reactors, or 72 reactor sites, why is it necessary to locate an additional 12 radioactive waste facilities, as is currently

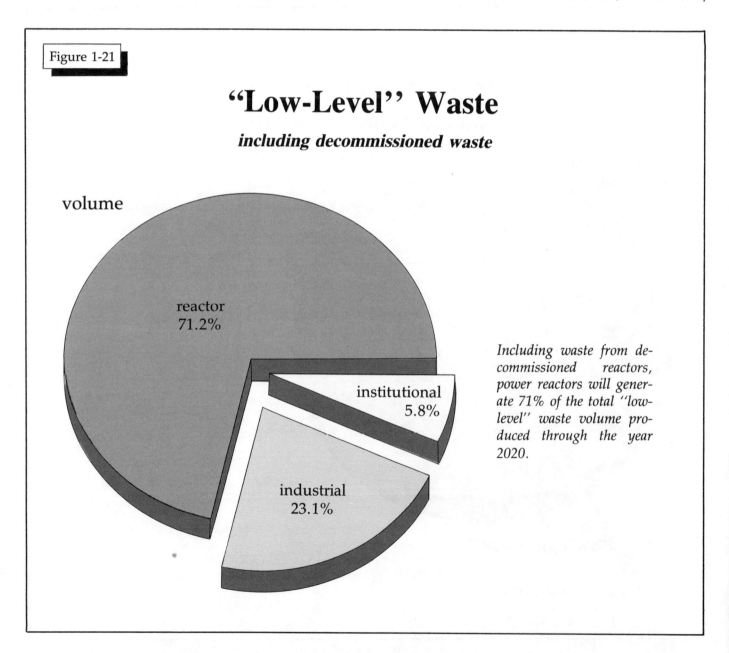

Figure 1-21

"Low-Level" Waste
including decommissioned waste

volume

reactor 71.2%

institutional 5.8%

industrial 23.1%

Including waste from decommissioned reactors, power reactors will generate 71% of the total "low-level" waste volume produced through the year 2020.

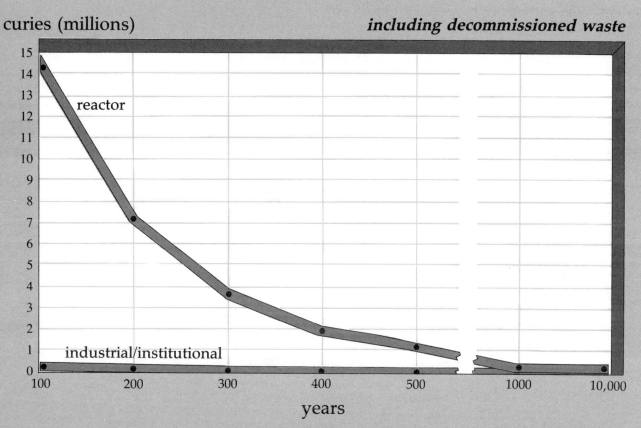

Figure 1-22

"Low-Level" Waste
Radioactivity as a Function of Time
including decommissioned waste

Waste from nuclear power reactors will remain radioactive for long periods of time. Even 10,000 years after generation, about 200,000 curies of long-lived radioactivity will remain from today's operating reactors.

being planned? Or stated another way, why move 99 percent of the radioactivity to a new, uncontaminated site, when 1 percent of the radioactivity could be moved to already contaminated reactors?

If additional waste were brought to and stored at reactor sites, a 40-year volume equal to about one-half the 40-year waste volume produced by each reactor, or 18,000 cubic meters, would have to be stored at the 115 power reactors. This additional storage at reactors would accommodate all waste generated in the U.S. to the year 2020.

But we're getting ahead of ourselves here. Let's bracket the issue of what will be done with this waste just a little bit longer, or at least until Chapters 4 and 5. First, we need to examine how "low-level" waste has been, and continues to be, "disposed of," and with what results.

CHAPTER 2
Experience at "Low-Level" Waste Landfills

*"Water, water, everywhere,
But not a drop to drink."*

CHAPTER 2
Experience at "Low-Level" Waste Landfills

The burial of commercial "low-level" waste in landfills began in the 1960's. Previously, the small amounts of such waste that had been generated were dumped at sea, along with much larger quantities of military radioactive waste. However, as production of waste from the fledgling commercial nuclear industry increased, waste generators complained about the cost of sea dumping. In response, the Atomic Energy Commission cleared the way for the burial of commercial waste on land (Barlett, 1985).

Initially, the Commission accepted commercial waste and buried it along with military waste at federal facilities in Oak Ridge, Tennessee and Idaho Falls, Idaho. But this temporary plan was soon followed by another arrangement: government-licensed, privately-managed dumping grounds. These private operations developed haphazardly, and not in accordance with any comprehensive federal plan.

The first facility was opened in 1962 near Beatty, Nevada. The following year, burial grounds were opened in Maxey Flats, Kentucky and West Valley, New York, and two years after that, another began operating in Richland, Washington. In 1967, a landfill opened at Sheffield, Illinois, while another facility opened at Barnwell, South Carolina in 1971.

Of the six commercial radioactive landfills which have operated in the United States, three are now closed because of serious problems: Maxey Flats, West Valley and Sheffield. All three have had water infiltration into trenches, subsidence of earthen trench covers and erosion. At each site, radioactivity has migrated and expensive remedial actions are continuing. Rather than being maintenance-free, stabilized landfills, these sites have ended up requiring active maintenance *within 10 years of trench closure*.

The problems at the three landfills in the humid North have been due to the following factors:

1) *Superficial geologic site characterizations.* In other words, sites were chosen without an adequate understanding of their geology and groundwater characteristics. In part, this was due to the lack of rigorous government siting criteria, and license and information requirements at the time. Contrary to the Nuclear Regulatory Commission claim that final siting decisions were

"based largely on hydrogeologic and economic factors (NRC,1980)," in fact, according to the Illinois Geological Survey, "the geologic descriptions were fairly superficial." The net result was that these sites were located in areas with large amounts of groundwater. One landfill, West Valley, was sited in a swamp, and another, Sheffield, was placed in an underground stream bed.

2) *Degradation of waste containers* and the waste itself, in a much shorter time than was anticipated.

3) *Subsidence, cracking and erosion of trench caps,* which has led to water infiltration into trenches. This has produced what is known as the "bathtub effect," where water enters a trench and is trapped, creating a "bathtub" full of water and radioactive waste.

4) *Underground lateral migration of water* through sand lenses and fractures in clay and shale.

5) *Presence of chelating agents.* Normally, these chemicals are used for cleaning radioactive contamination from pipes, etc. They work by turning radionuclides into heavy molecules that can't attach to anything else and are therefore easily moved (or washed away). However, when chelating agents are buried in landfills, they combine with normally insoluble radionuclides like americium, cesium, strontium, plutonium and cobalt, making them unable to bind to the surrounding soil and thereby allowing them to migrate out of waste trenches and even off-site.

6) *Lack of stabilization* and long-term care funds for the facilities. This has had the effect of passing the financial burden from waste generators and disposal companies to state and federal taxpayers, and to future generations.

And what of the three facilities that are still operating?

The Barnwell, South Carolina radioactive landfill is also located in a high rainfall area. Yet, it has not had a buildup of radioactive leachate because of the more porous clay and sand mixture upon which the site rests. This soil allows radioactive water to drain out and also permits it to evaporate during the hot summer months. Nevertheless, some radionuclide migration has taken place, and there is reason to believe the problem will get worse over time.

The other two operating sites, Beatty, Nevada and Richland, Washington, are both located in dry regions. They have apparently not experienced the same difficulties as the

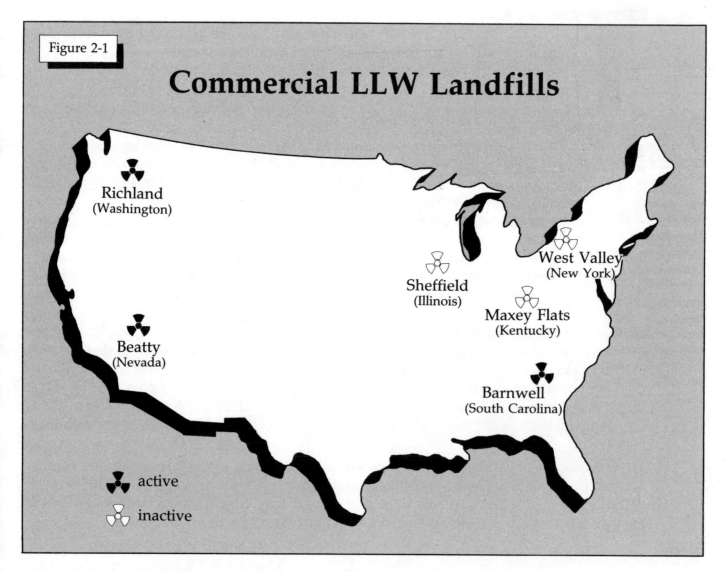

Figure 2-1

Commercial LLW Landfills

Richland (Washington) — active
Beatty (Nevada) — active
Sheffield (Illinois) — inactive
West Valley (New York) — inactive
Maxey Flats (Kentucky) — inactive
Barnwell (South Carolina) — active

sites in more humid areas, though their operation has by no means been completely trouble-free.

Figure 2-1 shows the location of the six commercial landfills which have operated or are still operating. A brief discussion of experiences at each of the six sites follows.

Maxey Flats

The Maxey Flats disposal site is located on a flat-topped ridge in northeastern Kentucky, about 65 miles east of Lexington. The area is quite humid, with average annual precipitation of 46 inches. The land was originally purchased in 1962 by the Nuclear Engineering Company (NECO), then sold to the state of Kentucky, which granted a lease to NECO. The facility operated from 1963 to 1977.

The 300-acre site includes a restricted access area of 42 acres, 25 of which contain "low-level" waste. Wastes were disposed of in 48 trenches, plus hot wells and special pits. The unlined trenches, carved out of dark blue to greenish shale and fine-grained sandstone, are 5 to 210 meters in length, 3 to 22 meters in width and 3 to 10 meters in depth (Mills,1983). The soil covers over the trenches are 0.3 to 5 meters thick. The site is cut by, and sits 300 to 400 feet above, tributaries of the Licking River, which feeds into the Ohio, the water supply for over one-half million people. An estimated 4.75 million cubic feet of radioactive materials were buried at Maxey Flats, with an estimated total radioactivity of 2.25 million curies, including 64 kilograms of plutonium.

The Maxey Flats site has experienced severe water infiltration problems. Because of the low permeability of the surrounding earth, water entering the more permeable trench caps accumulates and eventually fills the trenches, mixing with the burial materials and becoming radioactively contaminated. This is the so-called "bathtub effect." By 1972, over one million gallons of water had accumulated in the trenches.

In 1974, a state of Kentucky report found that plutonium at the site had moved hundreds of feet from where it had been buried (Barlett,1985;Lipschutz,1980). Plutonium, which has a half-life of 24,000 years, is one of the most toxic substances ever created.

The following year, the U.S. Environmental Protection Agency, following up on the state report, noted that although the burial site had been "expected to retain the buried plutonium for its hazardous lifetime," the plutonium had actually migrated from the site "in less than ten years." While U.S. Ecology, formerly called NECO, claimed that subsurface migration of plutonium was impossible, the EPA found plutonium approximately 3 feet deep in core drilling samples. Plutonium was also found in surface soil, monitoring wells and drainage streams. As the EPA report put it:

> If 100 percent retention of waste is the goal of shallow land disposal, continued burial of plutonium in humid climates using present waste forms, containers and trench construction methods will not achieve the goal (Meyer,1975).

As the trench caps subsided at Maxey Flats, about two million gallons of water per year began entering the trenches. In an effort to alleviate the problem, an evaporation system was installed in 1973. Trench water was pumped into tanks and evaporated, and the resulting sludges were stored in a tank on-site. However, since the evaporator processed only 1.2 million gallons at its maximum capacity, an excess of 800,000 gallons remained in the trenches and on-site tanks. To make matters worse, rather than containing radioactivity, the evaporator served to distribute radioactivity over the Kentucky countryside by releasing small quantities of tritium as steam.

In 1977, the detection of cobalt-60 and manganese-54 in a seep in *newly* opened trench 46 led to the closing of the Maxey Flats landfill. The state of Kentucky purchased the lease rights from NECO for $1.2 million, and now maintains the site at a yearly cost of $1 million. In addition, research and development costs are paid by federal agencies.

In November 1981, as a temporary remedial measure to halt water infiltration through trench covers, the state covered three-quarters of the trenches with 15 mil polyvinyl chloride. The polymer membrane was covered with a geotextile material for structural strength and 45 centimeters of earth to protect against weathering.

Analyses of water in the trenches (Dayal,1983) have shown that tritium and dissolved cobalt-60, cesium-137, strontium-90 and plutonium are present, though only tritium and plutonium have migrated underground appreciably. The chemical EDTA, a chelating agent, interacts with these normally insoluble radionuclides, making them soluble and permitting them to migrate. Even very small quantities of EDTA can have this effect. Levels as low as 2 parts per million result in 95 percent cobalt-60 in solution (Dayal,1983). Tritium is the most abundant radionuclide, with levels ranging up to 1830 times the off-site maximum permissible concentrations (MPC) allowed by the Nuclear Regulatory Commission. Following tritium, the most abundant radionuclide is strontium-90, with levels ranging up to 1000 times off-site maximum permissible concentrations.

Tritium has already been detected in the close-in test wells, clearly showing that it has migrated outside the restricted area. It has not been detected in the monitoring wells which ring the perimeter of the site, suggesting that off-site migration may not yet have taken place. However, this is not certain, since transport from the trenches takes place in fractures, and monitoring wells would have to be precisely located in fractures to detect radionuclide migration (Fisher,1983). According to the US Geological Survey, tritium is migrating at the rate of 50 feet per year (O'Donnell,1983).

Tritium has been detected in the sap of nearby maple trees, but it is not clear whether this is due to the evaporator or subsurface migration. Sampling of trees was done in March of 1983, several months after plastic covers were in place and the evaporator had been shut down. Tritium levels in the sap of maple trees ranged from 10,000 to 290,000 picocuries per liter. By comparison, trees growing 20 kilometers from the site have tritium levels on the order of 1,000 picocuries per liter (Rickard,1983).

The highest concentrations of tritium in tree sap, amounting to 10 percent of maximum permissible concentration, came from trees about 500 meters from the site. These were closest to a trench containing approximately 300,000 curies of tritium from Department of Energy nuclear warhead refurbishing operations at Mound Laboratory. An estimated 600,000 curies of tritium are buried at Maxey Flats.

Tritium, probably from the evaporator fallout, has also been detected in domestic well water. Further, milk samples from cows within 3.1 kilometers of the site have shown elevated tritium levels. Subsurface migration of radionuclides other than tritium has not been extensive compared to trench water concentrations.

Decommissioning plans for Maxey Flats were formulated by the state of Kentucky in 1984 (Kentucky,1984). According to those plans, 80,000 gallons of evaporator sludges stored in the tank farm would have to be solidified. To minimize further settling, the trench caps would be deep compacted, then covered with asphalt, cement or vegetation to prevent water infiltration to the trenches. Lateral groundwater flow would be eliminated with a cut-off trench about the perimeter of the site.

Excluding project management costs and contingencies, decommissioning cost estimates range from $52 million to $121 million. Westinghouse, the present site manager, has more recently estimated closure costs at $25 million. These figures assume that the land reverts to unrestricted use after 100 years. As of 1978, $250,000 had been put aside in a perpetual care fund (EGG,1983a). The state is attempting to locate waste generators in order to collect additional fees to pay closure costs.

If it indeed costs $121 million to decommission the site, this is equivalent to $60 per cubic foot of waste material buried there. It is interesting to compare this real cost to the ideal long-term care and closure stabilization costs of $2 per cubic foot quoted by a federal Energy Department contractor for a site of equivalent size (EGG,1983b). Obviously, the Department of Energy is not anticipating that past problems will be repeated at future radioactive landfills.

Sheffield

On October 2, 1966, the California Nuclear Company and the Illinois Department of Health signed a lease agreement permitting the establishment of a "low-level" radioactive waste disposal site near Sheffield, Illinois, in Bureau County, about 50 miles north of Peoria. The site is owned by the Department of Nuclear Safety on behalf of the state of Illinois, and, unlike the other commercial radioactive landfills in the United States, is directly licensed by the Nuclear Regulatory Commission.

The 20-acre site is in a region of abandoned coal mines. It is bordered to the north by rolling terrain and the 40-acre NECO hazardous waste disposal site, which is on the Environmental Protection Agency's Superfund list of the worst dumps in the country. Precipitation in the area averages 35 inches per year. The water table varies from 15 to 45 feet below the ground surface.

The burial of radioactive waste was first authorized on August 1, 1967. NECO, now called US Ecology, the same site operator as at Maxey Flats, Kentucky, succeeded the original lessee on March 19, 1968 (NRC,1981a).

No detailed inventory of trench contents at Sheffield exists. In approximately 10 years of operation, three million cubic feet of waste containing 60,000 curies of fission products, 55 kilograms of special nuclear material (plutonium and uranium-235) and 600,000 pounds of source material (uranium-238 and thorium), were buried in 21 trenches (NRC,1981a). The dismantled Elk River Reactor is also buried at Sheffield.

Wastes at Sheffield were dumped haphazardly to within 3 feet of the original land surface, after which trenches were backfilled and contoured to aid water runoff. The trenches range in length from 35 to 580 feet, with widths of 8 to 70 feet and depths 18 to 26 feet. The trench caps are 3 feet thick. The bottoms of all trenches were supposed to be 7 to 10 feet above the water table, but trench 18 actually intercepts the water table.

In 1976, when space in the present 20-acre site was exhausted, NECO applied for a 168-acre expansion of the site. The application was later withdrawn at the recommendation of the state attorney general, who vigorously opposed further expansion. The last burial occurred in April 1978. In 1979, NECO attempted to abandon Sheffield and the site was officially closed. The Nuclear Regulatory Commission obtained a court injunction, forcing NECO to remain on the site. The perpetual maintenance fund for the site has been exhausted, and the state attorney general has a $97 million suit pending against the operator, US Ecology. As of summer 1987, the state was on the verge of settling with the site operator, but would not divulge the terms.

The glacial deposits underlying the site have low permeability, but continuous pebbly sand strata of high permeability lie underneath 60 percent of the site. These strata constitute the principal pathway for groundwater flow off the site near the east boundary (Johnson,1983). The sand strata were previously thought to be discontinuous, but the early geological characterizations of the site were quite superficial (Cartwright,1982).

Tritium was first detected in a well in the spring of 1978, and thus far is the only radionuclide above background concentrations detected in monitoring wells (Healey,1983). In the three-year period between 1975 and 1978, tritium had moved 75 feet east, but the migration rate has since accelerated. On October 12, 1982, tritium had migrated 665 feet from the closest burial trench and was discharging into a pond on private property. The following January, NECO purchased 120 acres, including the pond, as a buffer zone. Radiation levels in the plume ranged up to 80,000 picocuries per liter. The tritium concentrations near the site vary between 200 and 2 million picocuries per liter (two-thirds of off-site MPC).

More recent chaser dye tests by the Illinois State Geological Survey indicate a narrow tritium plume, now thought to be an underground stream, moving east at rate of *2000 to 3000 feet per year* (Chinn,1984). In light of this recent information, the state is reassessing its position on the purchase of additional buffer acreage, and is, instead, investigating the possibility of the federal Energy Department assuming responsibility for perpetual care of the site. Local citizens have requested that the radioactive waste materials be exhumed and placed in above ground storage.

The trench caps at Sheffield have severely eroded and cracked, and they now require continual maintenance. Erosion occurs between the trenches, and cracking occurs at the sides of the trench caps. Some of this has been the result of numerous incidents of subsidence, generally following periods of heavy precipitation. Seventy-four separate incidents took place just between November 1978 and May 1979 (NRC,1981). Most are small depressions on the order of 3 feet in diameter, but some are as large as 15 feet wide and 10 feet deep. Problems have also been created by periods of prolonged dryness, which lead to trench cap cracking, erosion and water entering the trenches.

West Valley

The West Valley site is located 30 miles southeast of Buffalo, New York in rural Cattaraugus County. The 3,345-acre site was acquired by New York State in June 1961. It was leased to Nuclear Fuel Services (NFS) two years later for the construction of a nuclear fuel reprocessing plant to chemically separate plutonium from fission products (Resnikoff,1977). The site is located on a plateau bounded on three sides by bedrock hills. Buttermilk Creek cuts through the center of the site, about 200 feet below the plateau. The soil in the burial area consists of highly impermeable silty till, with up to 10 feet of more permeable glacial till overburden. The silty till contains numerous vertical fractures and layers of more permeable sandy strata. The average annual precipitation in the area is 41 inches.

Construction of the reprocessing plant began in 1963, and the plant started operating in April 1966. During the six-

Photo by Diane D'Arrigo

Waste drums sitting in a pool of water outside the West Valley, New York radioactive landfill.

year operating life of the Nuclear Fuel Services plant, 624 tons of irradiated fuel were processed and the resulting high-level wastes were placed in two underground tanks. NFS operated a 7-acre burial area, licensed by the Nuclear Regulatory Commission, for "solid" radioactive wastes from plant operations. These included highly contaminated fuel hardware, obsolete equipment, and degraded process solvents which were "absorbed on a suitable solid medium," that is, kitty litter (DOE,1978). Approximately 528,000 curies of extremely radioactive materials (volume, 139,000 cubic feet) were buried in 3 by 7 foot unlined holes, 50 feet deep (Battelle,1979). One-half ton of irradiated fuel elements were also buried in the NRC-licensed burial ground.

In 1963, the same year the Maxey Flats landfill went into operation, the New York Department of Health granted NFS a waiver of its prohibition against land burial. This was the New York State "license." The 25-acre burial plot, part of which is located in the former Spitler's Swamp, is adjacent to the NRC-licensed burial area. Both burial areas have the same topographical, surficial and hydrological features. Approximately 740,000 curies of radioactivity in a volume of 2.3 million cubic feet were buried in unlined trenches in 7 acres of the commercial landfill (NRC,1980). The commercial landfill ceased operation in 1975 when radioactive trench water began to seep through the trench caps.

The commercial radioactive landfill has two sets of trenches, northern and southern. The northern trenches, numbered one through seven, were cut and filled in the years 1963 to 1969. The southern trenches, numbered eight through fourteen, were cut and filled between 1969 and 1975. Within a few years of filling, water began infiltrating and accumulating in the northern trenches. The cause of this water infiltration was comparable to the Maxey Flats case—the trench caps were more permeable than the walls and bottoms of the trenches.

Considering trench number five as an example, New York State data (Cashman,1982) show that the water rose at a rate of one foot three inches per year in the period 1969 through 1971. In the period 1971 through 1973, water levels rose at an even faster rate of three feet three inches per year. Between 1973 and 1975, until contaminated trench water broke through the caps, the water levels rose at a rate of five feet per year. Each foot rise in the water level is approximately equal to 100,000 gallons of water. The radiation levels in this trench water ranged up to 1000 times off-site MPC. The trench cap breakthrough in trenches four and five actually exposed the waste materials to the open air. The covers were then repaired, and, in 1978, an additional 4-foot cover was added, contoured and seeded. The water infiltration rate then dropped to between 1 and 1.5 feet per year, the same rate as during the period 1969 through 1971. But water infiltration has not ceased, implying that infiltration may also be due to underground migration along sandy strata.

Based on the supposedly more successful experience with the southern trenches, covered with an 8-foot cap, federal agencies, in 1978, were quite optimistic about the future performance of the northern trenches. "It is too soon to determine the effectiveness of these procedures, but experience with the southern trenches would indicate that infiltration through the caps should now cease and erosion should be prevented (DOE,1978)." Indeed, the performance of the southern trenches, which began accepting waste in 1969, was initially better. Just looking at trench number 13, one of the last trenches constructed, the water levels remained constant for the years 1973 through 1977.

However, in the period 1977 through 1979, at the same time DOE was extolling the virtues of the 8-foot covered southern trenches, the water levels began to rise at a rate of two feet two inches per year (Cashman,1982). Because the summer of 1979 was dry, the clay trench covers may have cracked more than usual, thereby providing a water infiltration route the following winter and spring. In addition, a part of trench 13 is located in a former swamp.

The infiltration rate increased further in 1980 and the state began pumping the "improved" southern trenches to avoid trench cap breakthrough. The pumping operation, a process best called "controlled leakage," involves transferring the radioactively-contaminated water to a treatment facility, where the water is passed through ion-exchange resins and then released to the Cattaraugus Creek watershed. This treatment method releases small quantities of all the radionuclides contained in the water into the environment. The radioactive sludges are then buried in the NRC-licensed burial area.

Similar to Maxey Flats, the West Valley trenches contain high levels of the chelating agents EDTA and DPTA (NRC,1984c). The presence of these chelating agents will lead to more rapid migration of cobalt, plutonium, americium and strontium. The treatment facility has questionable effectiveness in the presence of these chemicals.

In spring 1987, water levels in trench 14 reached the point where the trench had to pumped down to avoid cap breakthrough. Trench 14 has an 8-foot clay cover, and was the last, and supposedly best-constructed, trench completed at West Valley. It is located in the area of the former Spitler's Swamp, and is also hydrologically connected to trench 13.

Clearly state geologists and commercial landfill operators placed too much emphasis on the impermeable character of clay, failing to take into consideration the presence of horizontal and vertical cracks in the clay, which provided routes for migration of radioactivity.

Another serious concern at both West Valley burial areas is rapid erosion of the plateau banks towards the burial ground area (NRC,1984d). In 1978, the steeper trench slopes were covered with a plastic sheet and a layer of crushed stone "to anchor and protect" them from sunlight (NRC,1980). This is clearly only a temporary solution. In an attempt to eliminate continual maintenance at the West Valley site, New York State, through its Energy Research and Development Authority, has contracted with the U.S. Department of Energy to study how to decommission the burial areas.

While much attention was focused on the state-licensed commercial burial ground, little notice was paid to the NRC-licensed burial area. Because of high tritium levels in the ravine between the state-licensed and NRC-licensed burial areas, an official of the New York Department of Health, in rare agreement with citizen groups, requested the exhumation of the NRC-licensed burial area (DOH,1979).

Potential problems with the NRC-licensed burial area would have gone unnoticed had the New York State Geological Survey not detected kerosene contaminated with radionuclides in a monitoring well in 1983. This plutonium and iodine-contaminated kerosene was detected 20 feet below ground, approximately 60 feet from the nearest potential source. The radioactive concentrations, on the order of 1,000 times the off-site maximum permissible concentrations, are due primarily to iodine-129, which has a half-life of 17 million years. Since federal agencies and the New York State Geological Survey had assured everyone that the silty till was impermeable, and that plutonium would migrate only one-quarter of an inch a year, this plutonium migration came as yet another unpleasant surprise.

West Valley has also had problems with chelating agents. Tributyl-n-phosphate (TBP) is a chelating agent used in the processing of nuclear fuel. After irradiated fuel is dissolved in nitric acid, this organic solvent is added to selectively remove uranium, plutonium and some fission products. Eventually this solvent becomes too contaminated and requires disposal. At West Valley, this solvent was buried in 1000-gallon tanks in the NRC-licensed burial area.

In this experiment taking place in the field, it is now believed that kerosene with TBP increases the permeability of silty till. A radiation plume, extending over 30 feet in width, with uncertain depth, is now moving towards a feeder stream of Buttermilk Creek, Erdman's Brook, to the north of the NRC-licensed burial area. A string of interceptor wells has been placed at the edge of the plateau, near Erdman's Brook. The NRC has funded a major research and development program (ORNL,1984) to determine how to stem the tide. At present, wells have been drilled near the likely leaking tanks and radioactive solvent is being removed as it accumulates.

Complicating the problem is the fact that kerosene

Photo by the Environmental Protection Agency

Waste drums sitting in water in a partially filled burial trench at West Valley, New York. After trenches were filled and mounded over with an 8-foot clay cover, water continued to enter the trench cavity.

floats on water, thereby rising and falling depending on the seasonal fluctuations. Fortunately, burial holes SH-10 and SH-11, the likely kerosene sources, contain very little radioactivity, less than 1 curie total. However, carbon steel tanks in holes SH-27, SH-28 and SH-29 contain about 1,100 curies in organic solvent. Since these carbon steel tanks will surely degrade, the Radioactive Waste Campaign, and local groups, such as the Coalition On West Valley Nuclear Wastes, have called for exhumation of these tanks *before* plutonium begins to migrate. Eight underground tanks in burial holes SH-10 and SH-11 were exhumed in fall 1986 at a cost of $1.25 million, demonstrating that exhumation of radioactive waste is at least possible.

As a result of Department of Energy efforts to solidify two underground tanks of liquid high-level waste remaining from earlier reprocessing operations, about 60,000 curies of "low-level" waste, primarily cesium-137, will be generated. DOE intends to put this waste, in the form of concrete within steel drums, into an above ground pyramid, called a tumulus (see Chapter 4).

Barnwell

The 280-acre Barnwell, South Carolina landfill is one of the main operating landfills for radioactive waste in the United States. In 1982, approximately 35,000 cubic meters, or 46 percent of the annual commercial radioactive waste volume, and about two-thirds of the radioactivity, were buried at Barnwell (EGG,1983c).

Earlier, Barnwell had accepted even larger amounts of waste. For example, in 1979, the Barnwell site accepted a full two-thirds of the nation's waste volume (NUS,1980). However, these quantities were reduced when the governor of South Carolina, sensing that the state was becoming the nation's dumping ground, imposed a monthly capacity limit of 3,000 cubic meters. Much of the excess now goes to the landfill at Richland, Washington.

The Barnwell facility is owned by the state of South Carolina, which has licensed Chem-Nuclear Systems, a subsidiary of Waste Management, Inc., to operate the site until 1992. The state has made clear its preference to close the site at that time, when another state in the Southeast will sup-

Photo by Chem-Nuclear

The "kick and roll method" of waste disposal, formerly practiced by Chem-Nuclear at the Barnwell, South Carolina radioactive landfill.

40 LIVING WITHOUT LANDFILLS

posedly open a facility.

The Barnwell site is located near the Savannah River, about 70 miles southwest of Columbia. Originally licensed in 1972, the site has a large capacity—2.4 million cubic meters. Radioactive waste is buried with the standard cut and fill method used at other sites. In the early 1970's, waste was buried quite haphazardly. Barrels were often rolled out of the rear of semi-trailers, in what has been called the "kick and roll method." At the present time, though, all waste materials must be solid, and they are packed tightly within the trenches in order to minimize void spaces.

The problems of package degradation, trench cover subsidence and water infiltration at Barnwell are no different from those at the three closed sites in the North, but the results have not been the same because of the nature of the surrounding medium. Unlike the Maxey Flats and West Valley sites, the surficial materials are more permeable sand and clay. Water entering the trench caps does not accumulate; rather, it passes directly through the bottom of the trench or evaporates during the summer months. Thus, unlike the Maxey Flats and West Valley landfills, radioactive materials do not "brew" in trench water at Barnwell. No trench water samples have yet been collected even though the mean annual precipitation is over 43 inches per year. Surface water runoff is also minimal because of the sandy soil and flat topography. Compared to the northern sites, little radioactivity has migrated out of the Barnwell landfill.

Nevertheless, some migration of tritium has taken place at Barnwell. In 1981, tritium was detected at the 21-meter depth, at levels of 116,000 picocuries per liter (Cahill,1983), indicating downward movement of water from the trenches. Tritium has also been detected above background levels as far as 75 meters southwest of buried waste (Cahill, 1983). The only other radionuclide to have migrated is cobalt-60, which was found in core samples beneath trench 2. The activity levels were 690 picocuries per gram of sediment, decreasing to non-detectable levels at depths 1.8 meters below the trench floor. Since the same chelating agent that migrated at the NRC-licensed burial ground at West Valley, tributyl-n-phosphate, was buried at Barnwell until 1982, we suspect that cobalt-60 and strontium-90 will begin to migrate as the waste packages degrade.

Richland/Beatty

The Richland, Washington, and Beatty, Nevada, sites are both located in extremely dry parts of the United States. The absence of precipitation greatly simplifies waste management procedures. The trenches remain open until completely full. In contrast to sites in the East, no attention is paid to eliminating void spaces, and trench caps are not compacted.

The 80-acre Beatty site is located near Death Valley, 105 miles northwest of Las Vegas. With annual precipitation of only four inches, the area is characterized as arid. The soil is sand and gravel, and the site sits on thick layers of clay. The water table in the area is 300 feet below the surface, which is many times deeper than the water tables at the three closed landfill sites in the East. In 1982, Beatty accepted 2 percent (by volume) of the nation's waste (EGG,1983c).

The 100-acre Richland site is located near the Oregon border, at the confluence of the Yakima and Columbia Rivers. The annual precipitation is six inches. The soil is sand and gravel. In 1982, Richland accepted 52 percent of the nation's waste, making it the main commercial "low-level" radioactive waste landfill for the United States.

Some important information on these sites has not been gathered. For example, no sampling of trench water is carried out. And according to a U.S. General Accounting Office report, major information *lacking* about the Hanford site adjacent to Richland includes, "rate of infiltration (the amount of water that is not evaporated and is free to move downward), rate and direction of groundwater movement, and interconnection between shallow and deep aquifers (GAO,1976)."

A serious disadvantage common to both Beatty and Richland is that they are located far from major waste generators in the East, thereby greatly increasing transportation costs and impacts. Both sites have also experienced minor problems with wind erosion and animal movement (e.g., gophers burrowing into trenches).

The Beatty operation has had problems involving radioactive drums being buried outside of the site boundary, and radioactive materials (e.g., tools) being removed from the site (most were later recovered). In addition, as is the case with the three closed-down facilities in the East, some shallow trenches at Beatty and Richland contain substantial amounts of extremely long-lived plutonium-contaminated waste (Barlett,1985). According to the GAO, "Through 1974 over 140 billion gallons of liquid waste containing about 5 million curies have been discharged into the ground at Savannah River, Idaho and Hanford (GAO,1976)."

The Verdict on Landfills

The generally poor results from the landfilling of radioactive waste are not altogether surprising. Radioactive landfills have much in common with toxic chemical and municipal landfills, which have an incredibly poor record. Barrels decay, water mixes with toxic chemicals and pollutes groundwater supplies. The Office of Technology Assessment estimates between 17,000 and 35,000 solid waste facilities are in need of cleanup, as many as 5,000 under Superfund legislation (OTA,1985). Cleanup cost estimates range from $10 billion (EPA) to $92 billion (National Audubon Society). The OTA estimates a cost of $60 million per site, including groundwater cleanup. According to OTA, none of the Superfund sites can be returned to formerly pristine conditions.

Sometimes, the cleanup of one Superfund site leads to the creation of another Superfund site. A case in point is the

Stringfellow Acid Pits site near Glen Avon, California. Between 1956 and 1972, over 30 million gallons of liquid hazardous waste were dumped there. Original geological studies wrongly concluded the site was on impermeable bedrock and that there would be no groundwater contamination. In fact, the site sits over the Chino Basin aquifer, a major source of drinking water for 500,000 people.

In 1972 groundwater contamination was wrongly interpreted to be the result of surface runoff, when, in fact, an underground contamination plume was moving towards the Chino Basin.

> Downgradient wells one mile and more from the site have revealed substantial contamination by toxic chemicals in concentrations sufficient for decertification of a drinking water supply. Alternate drinking water is being supplied to some local residents (OTA,1985).

Rather than remove all contaminated liquids and solids from the site, estimated in 1977 to cost $3.4 million, the state chose a lower cost option which involved removing contaminated liquids, neutralizing the soil, placing a clay cap, and installing interceptor wells to capture groundwater. Over four million gallons of contaminated water were disposed of at the BKK land disposal site in West Covina, which is now leaking and closed to hazardous waste. "The Casmalia Resources landfill that now receives 70,000 gallons per day from Stringfellow was recently fined by the EPA for inadequately monitoring the groundwater (OTA,1985)." About $15 million has been spent; the state estimates a permanent cleanup would cost $65 million.

The Office of Technology Assessment recommends that the toxic wastes at Stringfellow be excavated and stored on-site until the toxic waste can be rendered "as harmless as possible (OTA,1985)." According to OTA, "as time continues to pass, cleanup costs are likely to mount and cleanup may become infeasible if there is widespread contamination of more soil and groundwater (OTA,1985)."

The moral of the story is the same, whether we are talking about hazardous waste landfills or "low-level" radioactive waste landfills: they cannot keep what is dumped in them from getting into the environment. According to the EPA,

> At the present time, it is not technologically and institutionally possible to contain wastes forever or for the long time periods that may be necessary to allow adequate degradation to be achieved. Consequently, the regulation of hazardous waste land disposal must proceed from the assumption that migration of hazardous wastes and their by-products from a land disposal facility will inevitably occur (EPA,1981).

Dr. Peter Montague, Director of the Environmental Research Foundation has stated,

> We should probably stop thinking of so-called secure landfills as 'secure' places. I believe we would be better off as a society if we simply admitted that there is no such thing as a secure landfill. Landfills do not—in general—prevent landfilled wastes from entering the environment. Instead, landfills slow down the introduction of landfilled wastes into the environment. They retard (but do not reduce the ultimate quantity of) wastes flowing into the environment (Montague,1982).

CHAPTER 3

The Government Response
Too Little, Too Late

CHAPTER 3

The Government Response
Too Little, Too Late

Following the closure of three of the nation's six radioactive landfills in the second half of the 1970's, the three states with operating landfills began to lobby for a more equitable distribution of the "low-level" waste burden. After a series of transportation accidents, including one where a truck carrying radioactive materials caught fire (Shapiro,1981), two of the three remaining landfills, Beatty and Hanford, temporarily closed down, while the Barnwell facility squeezed incoming shipments to half the original volume, prompting a waste disposal crisis and sharply focusing attention on the "low-level" waste issue.

In Congress, legislation on "low-level" waste was attached to the Nuclear Waste Policy Bill of 1980. While the high-level waste portion of the bill was scuttled, not to be passed until 1982, the Low-Level Radioactive Waste Policy Act passed relatively unnoticed in the closing hours of the 1980 session. The act transfers enormous financial burdens and responsibilities from the federal government to the states, and strongly encourages the states to form regional groupings or compacts, new, decidedly undemocratic, forms of government.

Meanwhile, with public concern over the safety of waste disposal sites mounting, the Nuclear Regulatory Commission proposed new regulations designed to improve waste facility performance. As shown below, the proposed regulations were finally instituted in greatly weakened form. More recent NRC regulations, some proposed and others already instituted, have the potential to greatly increase the amount of radioactivity entering the environment. With landfill costs rising, the trend is to deregulate "low-level" waste by labelling it BRC or below regulatory concern. Another disturbing trend is the attempt to redefine some high-level waste as "low-level."

New NRC Regulations

The NRC released its siting regulations for radioactive landfills (NRC,1981b), 10 CFR Part 61, in July 1981. In their original form, the regulations were an important attempt by the agency to correct past problems. Unfortunately, these rules were never adopted. Instead, under nuclear industry pressure, the final regulations were greatly weakened, and attempts are being

Table 3-1

Change from *Proposed* to *Final* 10 CFR Part 61 Regulations

proposed regulations	final regulations	comments
Class B and C stability requirements—maintain structural stability within 5% under compressive load of 50 psi for 150 years	Class B and C stability requirements—maintain structural stability	NRC no longer requires 150-year time period, and other specific requirements
Liquids must be less than 1% volume	Same	
Waste disposal not within 100 year floodplain	Same	
Chelating agents greater than 0.1% not permitted	Applicant must show presence of chelating agents will still allow performance objectives to be met	Chelating agents at levels of 2 ppb will combine with cobalt-60 permitting the radionuclide to migrate
Applicant shall ensure sufficient funds available to carry out final site closure and post-closure surveillance and maintenance	Same	No specific requirement for sinking fund and surcharge on waste; no third party liability

Class C maximum limits[a] (Ci/m^3)

nickel-59	22	220[c]	Class C maximum limits on long-lived radionuclides increased by a factor of 10, but no new NRC calculations justify changes. Basis for changes: 1) markers will reduce possibility of intrusion, 2) difficulty of contacting C waste at 5 meters below surface
nickel-63	700	7,000[c]	
niobium-94	0.02	0.2[c]	
technetium-99	0.3	3	
iodine-129	0.008	0.08	
tritium[b]	108	no limit	
carbon-14	0.8	80[c]	
strontium-90	700	7,000	
cobalt-60	70,000	no limit	
cesium-137	4,600	same	
transuranic	10 nCi/g	100 nCi/g	
curium-242	no limit	20,000	

[a] *Concentrations above class C maximum limits mean material is unsuitable for surface disposal.*

[b] *These are minima for class C wastes. Final regulations allow any concentration of tritium-contaminated wastes to be considered as class B.*

[c] *This limit is for activated metals.*

made to weaken them still further. What started as a significant effort by the NRC staff to rectify past problems and reassure citizens became a virtually useless exercise.

Before discussing the watered-down regulations that went into effect in 1985, it would be useful to look more closely at the NRC regulations as originally proposed, and at their flaws. They were clearly a significant improvement over past regulations, and had overall siting objectives that were supportable: to identify sites that could be easily characterized and modeled, would have sufficient depth to the water table, would be well drained and would not be erosion-prone. To avoid cap subsidence, the regulations called for structural stability for the waste containers or waste forms, and minimization of void spaces. The regulations also required a buffer zone. Taken as a whole, the objective of the regulations was to keep water away from the waste, and to lengthen the travel time to the site boundary so that the dose to the public would be greatly reduced.

The proposed regulations, some of which may be found listed in Table 3-1, would have required that wastes retain their structural stability within 5 percent under a compressive load of 50 pounds per square inch, and that they be able to do so for 150 years. The regulations would also have mandated that free-standing liquids constitute less than 1 percent of the waste volume. Chelating agents, which combine and make mobile such radionuclides as cobalt-60, strontium-90 and plutonium, were not permitted in concentrations greater than 0.1 percent. Finally, the NRC regulations would have required that sufficient funds be available to carry out site closure and post-closure surveillance and maintenance. All these proposed regulations would have served to lessen past landfill problems.

The cornerstone of the 10 CFR Part 61 framework was the breaking of "low-level" waste into three classes—A, B and C—based on the types and concentrations of radionuclides contained in the waste, and on how the waste should be managed to minimize the risk of exposure to the public.

Class A waste, generally consisting of short-lived radionuclides, but also including some long-lived radionuclides, would be segregated but not stabilized in concrete blocks or other solid forms. Under this scheme, typical class A waste would be animal carcasses, trash and most research and medical waste.

Class B would include waste with higher concentrations of short-lived radionuclides than class A, and could also have some long-lived radionuclides. Being more hazardous, Class B waste would be segregated and placed in stable form or non-destructible containers. Filters and ion exchange resins from reactors would be in this class.

Class C would include waste with high concentrations of radionuclides, waste that is both hazardous and long-lived. It would have to be put into a stable form and segregated with a barrier to prevent inadvertent intrusion. Such a barrier could consist of an upper layer of class B waste or 5 meters of earth. Finally, greater than class C waste would be considered unacceptable for near-surface burial.

In developing the three waste classes, the NRC employed computer models of landfill leakage that included a large number of general assumptions about the physical and chemical form of the waste, the geology and hydrology of waste sites, and the sociology and psychology of how future generations might behave upon discovering radioactive waste. The NRC also assumed that waste sites would be guarded for a 100-year period after closure (called the institutional control period). Relying on hypothetical scenarios, the Commission derived specific radionuclide limits for each waste class. Simplified intruder scenarios, or simplified pathways to groundwater, for specific radionuclides from specific waste streams were looked at as a function of time. The NRC chose total doses of 500 millirems from direct irradiation, and 25 millirems per year from groundwater contamination, as its maximum acceptable exposure levels.

According to the Commission, people disturbing class A waste after the 100-year institutional control period would receive no more than a 500-millirem gamma exposure. This, said NRC, could be due, for example, to a farmer plowing a field where waste was buried, or to a construction crew digging the foundation of a building in such an area. The first is referred to as an intruder-agriculture scenario,

Table 3-2

Bone Dose Due To Group 4 Waste Streams in the Intruder/Discoverer Scenario

year	dose (rems)	dose without 2* of the streams (rems)
50	131.2	78.1
100	80.7	21.3
500	46.9	0.043

Group 4 waste streams are miscellaneous "special" or high activity streams: non-fuel reactor components, LWR decontamination resins, target sources, wastes from isotope and tritium production, and sealed sources. Data from Table 4.7, (NRC,1981d).

*Two waste streams, LWR decontamination resins and target sources, are removed from Group 4.

while the second is called an intruder-construction scenario.

With respect to class B waste, the NRC reasoned that the waste form would be recognizable into the future, and would thus be avoided when discovered. Therefore, argued the Commission, exposure would be limited to six hours. Under these conditions, a maximum dose of 500 millirems whole body dose could be incurred after the 100-year control period.

For class C, concentrations of specific radionuclides which would give rise to intruder or direct gamma doses less than 500 millirems, or groundwater doses less than 25 millirems per year, are shown in Table 3-1, Column 1. Of course, more restrictive levels of acceptability than those chosen by NRC, say 1 millirem per year, would lead to different radionuclide concentrations for the three classes.

Table 3-3

Classification of PWR Decommissioning Waste

category	burial volume(m³)*	total curies	class
Core Shroud	11	3,431,000	C+
Lower Grid Plate	14	553,000	C+
Thermal Shields	17	146,000	C+
Lower Core Barrel	91	651,000	C+
Lower Support Columns	3	10,000	C
Upper Core Grid Plate	14	24,300	C
Pressure Vessel Wall	108	19,200	A
Upper Core Barrel	6	1,000	B
Miscellaneous Internals	23	2,000	B
Lower Core Forging	31	2,500	B
Upper Support Columns	11	100	A
Guide Tubes	17	100	A
Bioshield Concrete	707	1,200	A
Upper Core Support Assembly	11	10	A
Reactor Cavity Liner	14	10	A
Vessel Head	57	10	A
Vessel Bottom	57	10	A

Adapted from (NRC,1978)

*Including shielded package, not just volume of contaminated component.

48 LIVING WITHOUT LANDFILLS

Particularly radioactive and long-lived are the so-called group 4 waste streams. These consist of miscellaneous "special" or high-activity streams, including reactor decontamination resins, target sources, non-fuel reactor components (control rods and activated reactor components), and wastes from isotope and tritium production, as noted in Chapter 1.

As seen in Table 3-2, column 2, doses from group 4 waste streams persist for 500 years and longer. The doses shown in Table 3-2 incorporate the NRC assumption that future discoverers of these 20th century relics would only be exposed for six hours. In other words, NRC assumes that people would recognize the waste and not disturb it extensively. Yet even 500 years after site closure, the dose is still 46.9 rems for a six-hour exposure. In contrast, background radiation doses are approximately 100 millirems per year.

Removing the two waste streams, ion exchange resins from decommissioning a nuclear reactor and target sources, lowers the dose considerably, particularly the long-term dose, from 46.9 rems to 0.043 rems, at 500 years. This clearly shows the benefit of removing these two class C + waste streams.

Similar benefits in reducing the direct long-term gamma dose are obtained by removing the class C components of decommissioned reactors. The contact dose from a pressurized water reactor shroud is about 3 rems per hour at 500 years after reactor operation (NRC,1978). With a calculated six-hour dose of 0.043 rems in the final environmental impact statement on Part 61 regulations, the high shroud dose is obviously masked by an NRC model which averages a small volume of hot components, such as pressurized water reactor shrouds, with a large volume of mildly contaminated components, such as concrete, leaving an average dose considerably reduced. The calculated dose thus depends critically on the model. For example, if, instead of staring at radioactive stainless steel, the future construction crew actually sold this metal for salvage and it was melted and recycled, the population and individual dose would become alarmingly high.

In Table 3-3, waste from a decommissioned pressurized water reactor is divided into class A, B, and C components. As the table shows, class C and greater components comprise only 150 cubic meters of the waste volume. Yet, as seen in Table 3-4, these class C wastes constitute 99.5 percent of the total radioactivity in a decommissioned pressurized water reactor.

Table 3-4

10 CFR Part 61 Classification of Decommissioned Reactor Waste*

reactor type	waste class	burial volume (m³)	radioactivity curies (thousands)	percent
BWR	C and C+	100	6540	99.8
BWR	A and B	128	13.4	0.2
PWR	C and C+	150	4815	99.5
PWR	A and B	1040	26.1	0.5

*Only neutron-activated wastes included; contaminated waste and radioactive trash excluded.

NRC Scenarios Unrealistic

While the scenario method of calculating future doses is reasonable, the particular situations chosen by NRC staff are clearly not the worst possible. Much more probable scenarios, involving higher individual and population doses, are easily conceivable.

As noted, the NRC based future low doses on the supposition that stable waste forms for class B and C wastes would be recognizable and fairly easily differentiated from dirt far into the future. According to the NRC, a construction crew or farmer, seeing these out-of-the-ordinary blocks, would halt operations and search records and deeds to determine the nature of these materials.

To see the unrealistic nature of this assumption, one need only imagine that if such a record search took place now, we would be examining deeds from before the arrival of Columbus! In the Northeast, most deed records date back to the Hudson Bay Company in the 1700's, but many are later yet.

A number of scenarios are perhaps more plausible than those used by the NRC.

Juarez Scenario

Most decommissioned internal reactor components are composed of stainless steel. This is a valuable resource that could be mined by future generations. A January 1984 incident in Juarez, Mexico (Science,1984) is just one example of how radioactive contamination could be distributed over a large population. In the Juarez incident, cobalt-60 was melted and recast into metal table legs and structural steel. The radioactive table pedestals were assembled into tables in Olivette, Missouri, to the west of St. Louis. Twenty houses and 4,000 tons of steel were contaminated. About 0.4 curies of cobalt-60 were involved in the accident.

All told, at least 200 people received gamma radiation doses ranging up to 50 rems. Four workers received 300 to 450-rem whole body doses. Two received hand and foot exposures on the order of 10,000 rems and developed wounds and blisters. NRC calculations do not take such a scenario into account.

Morocco Scenario

In Morocco, in March 1984, an iridium-192 radiation source used to radiograph welds at a fossil fuel power plant construction site, was taken home by a laborer (NRC,1985a). Unbeknownst to the worker, the gamma radiation source, about the size of a button and shiny like jewelry, was emitting extremely high levels of radiation: 150 rems per hour one foot from the source. Despite supposedly fail-safe security procedures, the loss of the source went undetected. Over a two-month period, eight members of the family, living in a one-room house, died of lung hemorrhages due to massive radiation exposures. In the NRC's waste classification system, this potent gamma radiation source would only be considered equivalent to class B "low-level" waste.

Stolen Tool/Pharoah's Tomb Scenario

Another plausible scenario not taken into account by the NRC staff is the possibility of radioactive tools being recovered and distributed. Such incidents have already taken place at radioactive landfills (Barlett,1985). At West Valley, New York, workers took radioactive tools from the former Nuclear Fuel Services reprocessing plant, and, instead of disposing of them, sold them at public auction four miles from the plant. A similar incident occurred at the Beatty, Nevada waste landfill. The presence of markers at the end of each burial trench did not prove a deterrent.

The Pharoah's Tomb Scenario presents another variant of this problem. Following the death of the Pharoahs, tombs were ransacked and valuables removed. Waste facilities will contain large amounts of radioactive stainless steel, waste to this generation, but potentially valuable to the next.

Love Canal Scenario

The NRC anticipates that after the 100-year institutional control period ends, a restrictive covenant in the deed would prohibit construction at waste sites and reserve land for light industry or a golf course. As the Love Canal tragedy shows, though, such restrictive covenants are insufficient to guarantee that future cancers and birth defects will not occur. At Love Canal, though no home construction took place on the landfill itself, toxic chemicals did seep into the basements of homes built around the landfill, causing miscarriages and cancers.

Manhattan Project facilities, just 40 to 45 years past the original mission, are in terrible disrepair, a situation that is just now being rectified by citizens forcing federal agencies, such as the Department of Energy, to properly maintain these sites. As is the case with most cemetaries, it is difficult to imagine waste facilities being cared for with the same zeal several generations removed from the original mission.

Under 10 CFR Part 61, after 100 years of institutional control, radioactive landfills could be unlicensed and uncontrolled. Almost any scenario is credible, particularly ones like those which have already occurred. The NRC has chosen low exposure scenarios and developed radioactive concentration limits which yield doses less than certain "acceptable" limits. But higher exposure scenarios and lower "acceptable" limits would lead to additional waste streams being classed as too hazardous for near-surface disposal. This is especially true for long-lived radioactive metals from decommissioned reactors.

The only fail-safe means of preventing future health effects is to remove long-lived wastes from the "low-level" waste stream and deal with them separately. Utilities are well aware that this increases the waste disposal costs for decommissioned reactors, as we show later. The bottom line is that future lives are being traded by utilities for reductions in present-day costs.

NRC Adopts Weakened Regulations

Flawed as the original A,B,C classification scheme was, under nuclear industry pressure the final rule, released December 1982, was weakened still further. The NRC altered the definitions of its "low-level" waste categories, allowing waste with more restrictive disposal requirements to move into less restrictive categories, thereby saving the nuclear industry money.

In going from its proposed "low-level" waste regulations (NRC,1981b) to the final version (NRC,1982a), maximum limits for the radionuclides strontium-90, technetium-99, iodine-129, nickel-59, nickel-63 and niobium-94, in class C waste, were all raised by a factor of 10. As shown in Table 3-1, for certain shorter-lived radionuclides (tritium and cobalt-60), no limiting values above class B exist.

The limits for alpha-emitting transuranics were relaxed by a factor of 10, from 10 nanocuries per gram (draft) to 100 nanocuries per gram (final), with materials having transuranic concentrations over 100 nanocuries per gram being considered unsuitable for surface landfills. With this change, and a corresponding change in Department of Energy regulations, several square miles of transuranic-contaminated earth at Hanford suddenly became just ordinary dirt. The final regulations have been strengthened in one respect—curium-242, which decays to plutonium-238, has been added to the table and is now covered by regulations.

In practical terms, these changes mean that greater amounts of radionuclides can be included in "low-level" waste, and the waste will still be considered class C, rather than greater than class C. This permits the waste to be managed without using (more expensive) isolation techniques. The regulatory changes also mean that decontamination resins are now considered to be class C, allowing them to be buried in landfills. Further, the changes reclassify certain reactor components from class C to class B. Since class C components must be layered or placed 5 meters below the surface, the easing of regulations translates into yet another cost savings for the utilities.

No new NRC calculations supported these changes. The only arguments advanced by the NRC (NRC,1982b) were that markers would reduce the possibility of intrusion (markers are less expensive than complying with regulations), and that wastes brought to the surface from 5 meters down would be diluted and provide a lesser dose than previously supposed. The Juarez incident and others point up the folly of these arguments.

Meanwhile, as indicated in Table 3-1, chelating agents are allowed in surface landfills, providing the license applicant can show that the performance objectives will still be met. Thus, the burden is on the landfill licensee to show that strontium-90, cobalt-60 and other radionuclides will not become mobile in the presence of decontamination resins containing chelating agents. The specific numerical requirement, 0.1 percent, was dropped.

In measuring the final 10 CFR Part 61 regulations against the history of trouble at radioactive landfills, it is apparent that they fail to deal adequately with the most fundamental issues. The financial assurance requirements in the final regulations are very weak, not specifically requiring a sinking fund or surcharge on waste, nor third party liability in case of radioactive leakage. The problem of inadequate long-term care funds at existing landfills was also not seriously addressed. While 10 CFR Part 61 does require financial assurance, it will be difficult for the NRC to enforce the regulation once a site is closed. Since it is entirely up to the private insurer whether insurance is provided at all, this is a very weak component of the new regulations.

Requiring class B and C waste to be in solid form addresses the matter of container degradation and subsidence, as well as the problem of trench cap cracking leading to water infiltration into trenches. But no specific numerical criteria are used, such as the length of time material must remain solid, the amount of degradation allowed under specific compressive loads, the percentage of chelating agents permitted, and so on. In the absence of specific criteria, the improvement can only be minimal. Whatever numerical criteria are called for are left to the licensing process for development, and citizens and the NRC staff have comparatively few resources with which to match industry in this process.

The high concentrations of long-lived radionuclides allowed into radioactive landfills by the final A,B,C classification scheme will pose a hazard for over 100,000 years.

NRC Defines the Problem Away

Below Regulatory Concern

The NRC's caving in to the nuclear industry didn't stop with the gutting of 10 CFR 61. More recently, the Commission has begun dealing with some waste by "defining" it away. As the cost of waste disposal has risen in the past few years, the NRC has accelerated its change of definitions to a dizzying pace.

Before March 11, 1981, all radioactive materials were buried at commercial radioactive landfills. With the closure of three commercial landfills in the late 1970's, and citizen resistance to new landfills, burial space became more scarce and expensive. Compared to utilities, institutional generators were less able to accommodate the increased costs.

The NRC came to the rescue of institutional generators by changing the regulations to allow waste with less than 0.05 microcuries per gram of tritium or carbon-14 to be disposed of down the drain (literally) or in municipal landfills (NRC,1981c). These wastes were redefined as being BRC, below regulatory concern. According to the NRC, the regulations were changed "because present regulations impose an economic and administrative burden on licensees that is not justified." Needless to say, the least expensive method of treating radioactive waste is to not treat it at all.

The process of defining away radioactive waste has continued with other generators and forms of radioactive waste as well. For example, the NRC permitted Duke Power to plow water cleansing sludges from the McGuire reactor into open land, about 1 acre per year. The NRC has also granted licensees permission to dispose of sludges and contaminated materials with low radioactive concentrations in quite "creative" ways. Contaminated sludges, soil and settling ponds have been land-farmed. Some contaminated materials have been sent to municipal landfills. Except for the HB Robinson 2 reactor, where up to 2 curies of settling pond sediment have been transferred to an on-site ash pond, the levels have been in the millicurie range (Branagan,1986).

Thus far, nearly 20 BRC requests have been granted. The number of requests and the amount of radioactivity is expected to grow as disposal costs at radioactive landfills rise. Under a congressional directive, the NRC recently established new rules for the "expeditious handling of petitions for rulemaking to exempt specific radioactive waste streams from disposal in a licensed low-level waste disposal facility (NRC,1986c)."

States and compacts may also deregulate "low-level" waste. For example, the Texas Department of Health recently adopted regulations allowing "low-level" waste with a half-life of less than 300 days to be buried in municipal landfills. Adoption of similar regulations on the federal level could reduce the amount of waste going to "low-level" waste facilities even further (Nucleonics,1987). In 1988, the Environmental Protection Agency is expected to promulgate new relaxed standards that could cut waste volume by 30 percent. According to Floyd Galpin, acting director of EPA's Criteria and Standards Division, "billions of dollars could be saved over the next 20 years through this (Nucleonics,1987)."

It's Magic—HLW Becomes LLW

In 1987, the NRC initiated proceedings on the issue of how to dispose of "low-level" wastes that are greater than class C, particularly reactor internals and control rods. While the final environmental impact statement for 10 CFR Part 61 labelled class C+ wastes as unsuitable for surface disposal, under utility pressure, the NRC may be having a change of heart.

Rules under consideration by NRC would change the definition of high-level waste (NRC,1987a), in effect moving waste from the high-level to the "low-level" waste category. This would save the nuclear industry enormous sums of money, since it is much more costly to dispose of waste in a repository than in a surface facility.

Under the proposed rule, high-level waste, formerly identified as irradiated fuel or the fission products from reprocessing irradiated fuel, would be classed according to its radionuclide composition. The sharp break between what was and was not high-level waste would be replaced with a sliding scale, allowing future revisions to slip more high-level waste into the "low-level" waste category. Waste previously unacceptable for surface facilities would be magically redefined as "low-level" waste. For example, under the proposed regulations, irradiated fuel could be disassembled. Under the regulations, the fuel rods themselves could be classed as high-level waste, but the activated stainless steel hardware, such as spacer grids, tie rods and metal end fittings, containing extremely long-lived activation products, could be classed as "low-level" waste. Many other waste streams would be subject to redefinition on a case by case basis. High-level waste in tanks at Hanford could also be reclassified as "low-level" waste.

Taken as a whole, the net effect of all these NRC actions would be that more short-lived waste would be poured down the drain, more long-lived waste would remain at the surface of the earth, and less long-lived waste would be put into long-term isolation facilities such as a deep underground repository.

Federal Legislation
Congress Drops the Hot Potato

The Low-Level Radioactive Waste Policy Act sailed through Congress in December of 1980. Given the vagaries and intricacies of the legislative process, the ease with which the bill navigated through the House and Senate might seem surprising, especially considering the usually controversial nature of nuclear issues. However, even a cursory glance at the act suggests why it was passed so quickly by Republicans and Democrats alike: its objective, quite simply, is to turn the problem of managing the nation's "low-level" radioactive waste over to the states.

The Low-Level Radioactive Waste Policy Act of 1980 (P.L.-573), though describing itself as setting forth "a federal policy for the disposal of 'low-level' radioactive wastes," actually makes each of the 50 states responsible for ensuring disposal capacity for the commercial "low-level" waste generated inside its borders. States may do this by themselves, or by joining with other states in compacts and establishing regional waste facilities. These compacts are to be congressionally approved.

The act contains only the most general guidance on achieving the objectives it mandates. It contains only a single deadline—January 1986—the point at which compacts were supposed to be permitted to restrict the use of their facilities to member states.

The legislation sets aside interstate commerce requirements of the Constitution for states joining in regional groupings (compacts). Normally the clause prevents states from impeding interstate commerce, and thus would allow one state to dump its wastes in another state. Under the Low-Level Radioactive Waste Policy Act, however, compacts may restrict their facilities to member states.

The act also has far-reaching implications for democratic governance. The commissions which administer compacts are made up of appointed officials. In most cases, state laws can be overridden by these non-accountable

52 LIVING WITHOUT LANDFILLS

compact commissions.

The federal legislation and compact agreements themselves impose substantial potential liabilities upon host states. For example, insurance coverage from private insurers is entirely voluntary. If private coverage is not available, citizens in host states would be liable. If some host states choose inexpensive landfills over more expensive above ground vaults, generators may attempt to send waste to the less expensive region, implying a possible revenue shortfall in a host state which chooses a safer, but more expensive technology. Federal consent to compacts extends only five years, though a waste facility may operate for 20 years, and a host state will have responsibility for many thousands of years. This creates tremendous uncertainty. Governor James Blanchard of Michigan has called for a revision of the federal legislation to guarantee more certainty for host states (Blanchard,1987).

Although several compacts were introduced for congressional approval before the January 1986 deadline, the majority of waste-producing states, with the majority of federal legislators, had not formed compacts in the intervening five years. With these states facing a waste disposal cutoff from the only three operating landfill states, Congress passed amendments to the Low-Level Radioactive Waste Policy Act.

The 1985 Amendments

Unlike the original act, the new Low-Level Radioactive Waste Policy Act Amendments of 1985 set specific timelines and penalties for inaction, while extending to January 1993, or 1996, depending on your interpretation of the law, the point at which operating facilities could bar waste from non-participating states. Specifically, the amendments allow existing landfills to set volume caps, and they condition continuing access to existing waste facilities through 1993 on the meeting of a set of milestones (Helminski,1986).

January 1988 Compacts must select a host state, and the selected host or individual non-member state must have a siting plan that sets a schedule for siting and license application. Most states interpret this federal requirement as state siting legislation, which has been introduced in many state legislatures. Compacts and states failing to meet the January 1988 deadline are charged twice the $20 base surcharge for each cubic foot of waste disposed of in South Carolina, Nevada and Washington. Twenty-five percent of this surcharge is collected by the Department of Energy and rebated to the states or generators depending on what additional milestones are met.

July 1988 Generators in states still not in compliance with the January 1988 deadline are charged four times the base rate per cubic foot.

January 1989 States still not in compliance with the January 1988 deadline can be denied access to the three operating waste facilities.

January 1990 A license application must be filed, or the governor must provide written certification to the NRC that the state will provide storage or disposal capacity beginning January 1993.

January 1992 The governors of states with operating landfills may assess generators from states which have not yet filed a license application a penalty of up to $120 per cubic foot for continued access to their facilities.

January 1993 If a state or compact is unable to provide disposal capacity, the state must, upon request of the generator, take title and possession and assume all liability for the generator's waste, or, if the state does not take title, 25 percent of the generator surcharges are to be rebated to the generators in that state. States with operating landfills are allowed to bar waste imports. States or compact regions without an operating waste facility are liable for all damages directly or indirectly incurred.

January 1996 This milestone is the same as for January 1993, except that the state automatically takes title and possession of waste, and liability for damages incurred.

States or compacts which achieve the milestones receive a 25 percent rebate of the disposal surcharges they pay to use the three operating sites allowed under the act. These rebates are to be used by states or compacts for waste disposal site development. If a state elects not to take title to the waste, the 25 percent rebate is paid to the generators in that state or region.

January 1996 is really "drop dead" time for states which refuse to accept responsibility for waste. The federal law says states *must* accept responsibility. Whether the federal government, under the Constitution, can, with the stroke of a pen, assign millions of dollars of responsibility to state governments is a legal question that is beyond the scope of this book. The Atomic Energy Act was created under the national security provisions of the Constitution, but it is a valid question whether the federal government can directly assign national defense costs to each state by an act of Congress. Similarly, can the federal government directly assign the costs of waste disposal, formerly a federal responsibility under the national security provisions of the Constitution, to the states? Debate over these issues is expected to heat up in 1988 as compacts choose host states.

The Low-Level Radioactive Waste Policy Act Amendments also direct the Nuclear Regulatory Commission to establish below regulatory concern standards for waste that need not be disposed of in a "low-level" waste facility (see page 51), and to provide "technical guidance" on alternative disposal methods. Mixed radioactive and toxic chemical waste is to be jointly regulated by the NRC and the EPA.

While increasing amounts of waste continue to be generated, compact regions are now in the frantic process of enacting siting legislation and identifying host states. The present regional groupings are shown in Figure 3-1. As can be seen, several states are still unaffiliated. Further, several compacts have become quite shaky as they have come closer to selecting host states. As a general rule, when a state within a compact is on the verge of being selected, state legislation is introduced to withdraw that state from the compact. This is true for North Carolina in the Southeast Compact, as well as Arkansas, Kansas and Nebraska in the Central States Compact. Such a move is also being considered by Michigan in the Midwest Compact. In the Central Midwest Compact, Illinois has agreed to be the host state, while in the Appalachian Compact, Pennsylvania has agreed to serve as host state.

Some compacts, such as the Midwest and Northeast, and some states, like New York, are developing so-called "incentive" packages to encourage a state or locality to volunteer to be a host area for waste. States and compact regions have also developed "public education" programs to reassure citizens and inform them of the "minimal risks" involved. Whether propaganda and bribes will do the trick remains to be seen, but the bottom line is, as it has been since 1972, that no new waste disposal facilities have yet

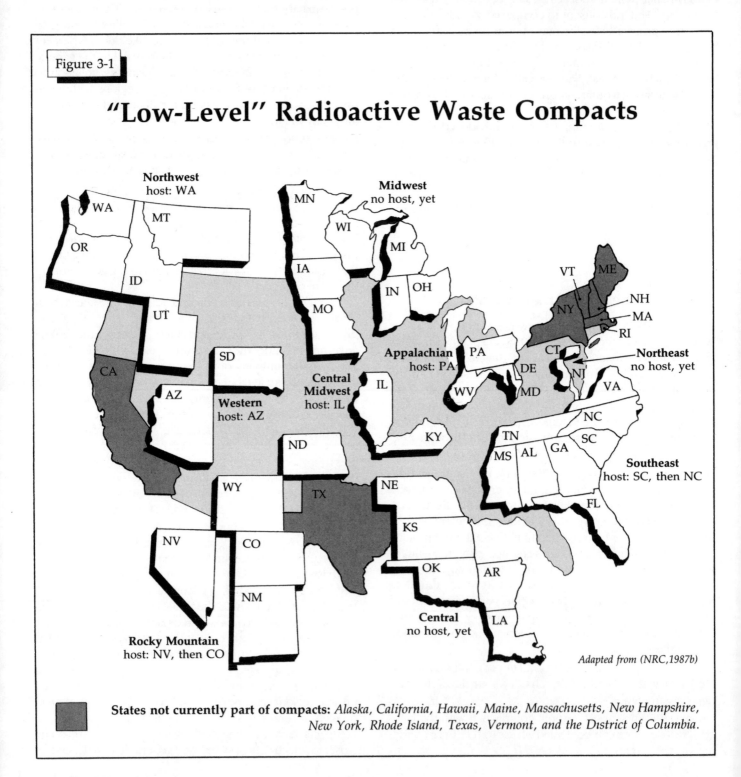

Figure 3-1

"Low-Level" Radioactive Waste Compacts

Northwest — host: WA
Midwest — no host, yet
Appalachian — host: PA
Northeast — no host, yet
Central Midwest — host: IL
Western — host: AZ
Southeast — host: SC, then NC
Central — no host, yet
Rocky Mountain — host: NV, then CO

States not currently part of compacts: Alaska, California, Hawaii, Maine, Massachusetts, New Hampshire, New York, Rhode Island, Texas, Vermont, and the District of Columbia.

Adapted from (NRC, 1987b)

been put into operation.

Citizen concern centers on several key issues:

Health and environmental effects. The primary concern is that leakage of radioactive materials could cause cancers and genetic effects. Citizens distrust health studies done by federal agencies which have a vested interest in making health effects seem minimal. Transportation to waste facilities extends the concern to citizens along transportation routes. Though the amount of radioactivity in each package is far less than in a high-level waste shipment, the packaging is much less rugged.

Property values. Since waste is an undesirable product, property values in an area near a waste facility will likely decline. The incentive packages are intended to offset this loss.

Equity. The benefits and risks of waste generation are distributed unequally. For example, electricity goes to major metropolitan areas, but nuclear waste is destined for rural areas. Since radioactive waste is also long-lived, inter-generational inequities are inevitable.

Industry track record. Predictions and performance by the nuclear industry have been notoriously poor. Rather than being given simple reassurances, stated with the same confidence as NASA before the Challenger disaster, citizens want to be told the truth by officials and agencies dedicated to protecting the public health. Citizens rightly distrust the nuclear industry and its emissaries. Will landfills leak? Do computer models accurately predict health effects? What exactly are the health effects? Have objective and protective health standards been set?

Public participation. Citizens want a meaningful decision-making role. Ever-increasing amounts of radioactive waste are being generated, beyond citizen control, and powerless rural communities are being asked to accept these wastes. Compacts, which can override state laws, are administered by gubernatorial appointees, and are not directly accountable to citizens.

Hazardous life. Certain radioactive wastes have hazardous lives of tens of thousands of years, yet the period during which states monitor and maintain waste facilities is limited to 100 years, the institutional control period. How trustworthy are long-term contracts, promises, and insurance and liability arrangements? Citizens are aware that when these promises and contracts expire, the hazard will still remain.

CHAPTER 4
Alternatives To Radioactive Landfills

CHAPTER 4

Alternatives To Radioactive Landfills

Myths are hard to break, and the most difficult is the disposal myth. Waste is supposedly "disposed of" in a landfill, but only "stored" in a warehouse or above ground storage vault. But really, waste in landfills is never "disposed of," if "to dispose of" means "to get rid of it" in some permanent way. The earth surrounding a landfill should be viewed as a container, an imperfect, never completely characterized, never entirely predictable container. And as Chapter 2 points out, landfills eventually release their contents into the environment.

The basic question is, can "low-level" waste be safely managed without landfills? Our conclusion is, yes. In this Chapter we discuss the general principles of any effective "low-level" waste management program: conservation, volume reduction, separation of waste according to half-life, and matching management options to the hazardous life of the waste.

Conservation

In general, waste resulting from any production process can be reduced through less production, product substitution, by recycling and by process controls. For radioactive waste, these methods will also serve to reduce the number of curies and the volume. Reducing the curie content is obviously the most desirable option, since this reduces overall toxicity as well. And, of course, smaller volumes imply less waste material must be managed.

Since this option also reduces waste disposal costs, generators have made great strides in reducing volume, as we describe shortly. But generators have also taken another approach as the price of waste disposal has risen: they have pressured regulators to change the definition of radioactive waste, calling some waste BRC, below regulatory concern. This really amounts to "legalized dumping."

Since almost 99 percent of radioactivity is produced at power reactors, an obvious first step in reducing waste is to cease operation of power reactors. If the sewer line were stopped up, it would make no sense to keep flushing waste down the toilet, but this is what is being done in nuclear power generation. Each day's operation produces many additional curies of "low-level" and high-level waste. Studies by Environmental Action Foundation (Nogee,1986) and Public Citizen (Kriesberg,1986), both public interest organizations, show that there is sufficient electrical overcapacity in the United States to phase out all nuclear reactors, with remaining excess capacity of over 15 percent. Numerous cost-effective alternative sources for electricity generation are available and are increasingly being employed: conservation, co-generation, photoelectric cells, small-scale hydroelectric, windmills, and fossil fuels. The cost of photoelectric cells has dropped precipitously (Sullivan,1986) at the same time as nuclear costs have escalated. Of course, even if nuclear power reactors are phased out, the waste already generated must still be managed.

Is every use of radioisotopes by industry, research institutions and hospitals absolutely necessary? Would citizens who live near proposed waste facilities accept the products if they came with waste? Would people accept a product if it came with a several generation commitment to waste management? It's beyond the scope of this book to investigate each product and the corresponding waste stream, and to pass judgment on its societal acceptability. However, it is worth reiterating that, as shown in Chapter 1, medical and research waste accounts for only a minute portion of all radioactivity generated, and much of that is tritium, with a short 12.3-year half-life.

Product substitution is entirely possible for smoke detectors, where americium-241 can be replaced with photoelectric detectors. Product substitution is clearly possible and preferable to food irradiation, since refrigeration and other means have allowed industrialized societies to flourish for generations. Do highway signs have to be coated with tritium, or will reflectors suffice?

Management practices can also reduce waste generation. At nuclear reactors, contaminated tools now remain in radiologically-controlled areas, where they are reused. An important management practice which reduces the curie content of "low-level" waste at power reactors is to ensure that new nuclear fuel is not defective, and to control the rate of power growth during start-up. This lowers the number of cladding defects, reducing the contamination which must be removed from reactor coolant. With this practice, the total quantity of radioactivity remains the same. It's just that the radioactivity stays with the fuel as high-level waste, rather than becoming more voluminous "low-level" waste.

Since radioactivity decays with time, holding waste for a time at the point of generation, approximately 10 half-lives, means waste will become non-hazardous over time. Because of the escalating cost of waste disposal, storage for decay is now a common practice among waste generators, particularly hospital and research institutions.

Volume Reduction

As a general rule, smaller volumes of waste are easier to manage than larger waste volumes. Reducing the waste volume also more efficiently uses the available space. Three volume reduction methods are being actively pursued by waste generators: compaction, incineration and redefinition of radioactive waste.

Disposal costs have risen enormously in the past 10 years due to increased disposal charges and state and local taxes. Disposal charges at the three closed landfills never fully reflected the true costs. Though high at the time, the charge by the state of Kentucky of $0.10 per cubic foot is a joke compared to estimated closure costs, which range up to $60 a cubic foot (Kentucky,1984).

Unfortunately, in the cases of Maxey Flats and West Valley, it is now the taxpayer, not the waste generator or site operator, who will foot the ultimate bill. According to the NRC, to exhume and relocate waste from a "low-level" waste burial ground would cost an estimated $1.4 billion (in 1978 dollars), and would take 21 to 25 years to accomplish (NRC,1980). And, this assumes that a substitute location could be found. Economic analyses for future landfills, by EG&G (EGG,1983b) for example, assume that landfills will operate perfectly and don't factor in major repair and maintenance costs.

Increased disposal costs, which more accurately reflect the true costs of waste disposal, have had some beneficial effects. Much more thought is being given to waste generation, since waste disposal charges are primarily based on volume. And volume reduction methods, particularly, supercompaction, have become cost-effective.

Unfortunately, incineration is also being promoted. This practice will disperse radionuclides into the environment and engender more opposition than a potential landfill. In addition, higher disposal costs have led to increased pressure to raise the levels below which waste is not regulated, which means that more waste will be disposed of by dilution. Of course, dilution does not cause the radiation to disappear. Each person exposed may receive a lower radiation dose, but dilution generally increases the total number of persons who ingest radioactivity and are exposed. The total dose and total number of health effects must still be calculated.

Compaction

One method of reducing the waste volume is to mechanically compress "low-level" waste under great force. This method is effective for some waste streams, and less effective on others. Compaction works best on compactible trash—that is, paper, cloth and plastics—from medical and research institutions, and from nuclear reactors.

Standard compactors, which develop forces of 10 to 15 tons, can reduce the volume of compactible trash by a factor of two to five. By first shredding waste, the volume reduction ratio nears five (EPRI,1982). Supercompactors can apply forces on the order of 1500 tons, compressing volumes by a factor of eight to nine. Newer designs, which involve shredding and forces up to 2200 tons (Baudisch,1985) have achieved volume reduction ratios up to 13.

Supercompactors can also reduce the volume of trash which is ordinarily non-compactible, such as power reactor hardware, radioactive gauges or devices, mop handles, wood, and so on. Other materials, such as liquids (biowaste, scintillation liquids, and contaminated oil) cannot practically be compressed.

The exact volume reduction which can be achieved depends on the specific waste streams accepted by the regional or state waste facility. In the Midwest Compact, the estimated reduction over all "low-level" waste streams, excluding decommissioning waste, is 21 percent (Midwest,1986a).

Since supercompactors can handle annual waste volumes on the order of 6,000 cubic meters per year, most states or regions will not require more than one machine. A portable supercompactor could reduce waste volumes before waste is transported, thereby lowering transportation costs. Since many utilities have purchased supercompactors, perhaps these can be employed by other waste generators.

Because radioactive gases and liquids are expelled when waste is compressed by a supercompactor, filters and collection systems are still required. Waste handling will also increase the risk of occupational exposure. Therefore, careful attention will have to be paid to protecting workers.

Below Regulatory Concern

This method, redefining waste as below regulatory concern, or BRC, is obviously the least expensive alternative for reducing the waste volume. Waste that is ruled BRC is not regulated. It can be disposed of down drains, buried in municipal landfills, or landfarmed at nuclear reactor sites. While approved levels have been low, in the millicurie range (Branagan,1986), there is great concern that the levels and quantities will rise under the recent encouragement of the NRC, which has streamlined the process for license approvals (NRC,1986c). For a more extensive discussion of BRC, see page 51.

Incineration

The idea behind incineration is simple enough—just as wood ash is much smaller than wood, radioactive ash occupies less volume than untreated waste. But unlike wood, the "smoke" and ash from "low-level" waste are both radioactive, and, anything touched by radioactivity becomes radioactively contaminated. While incineration reduces the volume of the original waste, it creates new waste—radioactive filters and scrubber solutions used to partially clean the incinerator exhaust gas.

Incinerators are particularly useful for oxidizing biowaste which would otherwise decompose in a landfill or storage facility, releasing 100 percent of the radioactive tritium and carbon-14 introduced to the incinerator. Of course, these two radionuclides would also be released quantitatively from an incinerator as radioactive steam and carbon dioxide. Contaminated oils and scintillation liquids, such as toluene and xylene, are also combustible and therefore are candidates for incineration.

While incineration greatly reduces the volume of paper and other combustible materials, the actual reduction factor is not as great as expected because the scrubber solution, which cleans the off-gas, must be solidified, and the ash itself must be put into a non-leachable form. While the NRC regulations do not require class A waste to be structurally stable, it is likely that all waste facilities will require incinerator ash to be solidified so that it cannot be dissolved and dispersed by water. Not even taking this into account, the Midwest Compact estimates a reduction of 26 percent over all "low-level" waste streams, close to the 21 percent estimate for a supercompactor (Midwest,1986a). Compacted waste may not satisfy the structural stability requirements of 10 CFR Part 61 since it may degrade or decompose in the presence of moisture. However, compacted waste is more structurally stable than ash.

One type of incinerator proposed as a centralized regional facility in North Carolina and Pennsylvania is a dual-chamber controlled-air incinerator (see Figure 4-1). Material is burned in the first chamber in a non-turbulent, deficient air process, at 1600° F, much like burning charcoal. The gases and particulates produced in the primary chamber are then passed into another chamber. This secondary chamber has an excess of oxygen to allow a complete burnup of waste gas and a better control of the combustion variables. The waste is burned in the secondary chamber at a temperature of 2200° F. This type of incinerator works well for the incineration of toxic materials, but not as well for radioactive substances. Off-gas treatment includes high efficiency particulate (HEPA) filters and a Venturi scrubber to control particulates, and charcoal adsorbers to capture iodine.

Incineration has four major drawbacks:

(a) Radioactive gases, such as radioactive steam and carbon dioxide, are released. Particularly important to contain is iodine, which can cause thyroid cancer. Since tritium acts like water, it can incorporate itself throughout the human body. This can cause DNA mutations, birth defects and cancer. Other radionuclides have to be contained by filters and scrubbers, adding to the waste volume.

(b) Burning of polyvinyl chloride plastic creates hydrochloric acid which must be neutralized. Hydrochloric acid emissions are one cause of acid rain. The more polyvinyl chloride that is burned, the greater the quantity of scrubber solution needed to neutralize the acid, contributing to the waste volume.

In addition, chlorine and oxidized paper are the precursors for dioxin formation. Dioxin has been detected at many incinerators (Doyle,1985). Some dioxin molecules are extremely toxic (Tschirley,1986). TCDD, a byproduct in the manufacture of Agent Orange, killed half the guinea pig test population at levels of only 0.6 micrograms per kilogram of body weight.

The exact formation mechanism of dioxin in an incinerator is unclear. Simply compressing polyvinyl chloride in a supercompactor, rather than incinerating the plastic, elim-

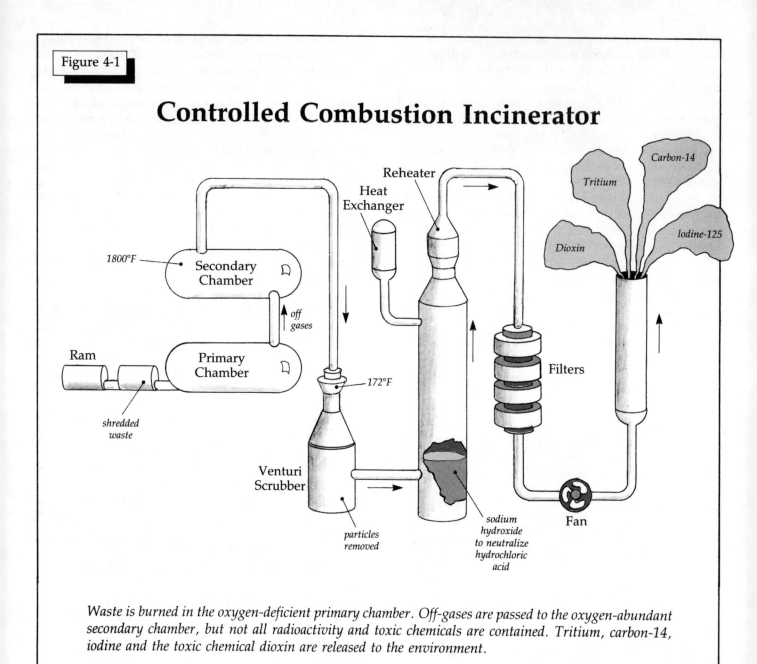

Waste is burned in the oxygen-deficient primary chamber. Off-gases are passed to the oxygen-abundant secondary chamber, but not all radioactivity and toxic chemicals are contained. Tritium, carbon-14, iodine and the toxic chemical dioxin are released to the environment.

inates this potential problem. Fortunately, institutions are increasingly replacing polyvinyl chloride with polyethylene.

(c) Incineration is more costly than supercompactors due to higher capital and operating costs. Economic pressures may force incinerator operators to cut corners by not solidifying radioactive ash or scrubber solution.

(d) Numerous accidents, such as explosions and fires, have occurred in incinerators, causing destruction of pollution control equipment and dispersal of radioactivity. For example, on June 10, 1983, a fire in the off-gas ventilation system at Nuclear Fuel Services in Erwin, Tennessee, caused extensive damage to the incinerator. Flames shot out of the chimney, burning up the radioactivity-containing filters.

Automatic shutdown mechanisms failed to operate. In January 1978, an explosion destroyed the incinerator at General Electric's Wilmington, North Carolina fuel fabrication plant (RWC,1985a).

Radioactive waste incinerators have been strongly opposed by citizens because of the dioxin threat and because these facilities release radioactive gases into the environment.

Separation of Waste by Half-Life

The essence of a safe and effective waste management program is storage for decay, since, unlike some toxic chemicals, radioactive materials become less toxic over time. This necessitates separating waste streams by half-life and designing the containment to match the hazardous life of the waste stream. This reduces the waste volume which must be guarded and maintained far into the future. It makes little sense to keep tritium waste, which has a half-life of 12.3 years, in the same containers as non-fuel reactor components, which contain extremely long-lived nickel-59 (80,000-year half-life).

After reducing curie production to an absolute minimum, by substitution of non-radioactive processes, waste streams containing short and long-lived radionuclides should be separated. As shown in Figure 4-2, radioactive waste streams should be separated by hazardous life into three categories: 100-year, 300-year and 10,000 + -year. The hazardous life of waste should be matched to the institutional control period and engineering capabilities of the storage system. Thus, a 100-year facility should contain waste which remains hazardous for no longer than 100 years. Waste containing short-lived radionuclides should be stored in engineered above ground structures until it decays to non-hazardous levels. Facilities for 100-year waste will contain waste from over 99 percent of waste generators. Because space in engineered structures is expensive, waste should be volume-reduced by supercompaction.

What exactly is hazardous, like what should be below regulatory concern, is difficult to define, but we would correlate "hazardous" with NRC regulations 10 CFR Part 20, as shown in Chapter 1. Properly, these regulations relate to hazard in water or air, but much thought has gone into setting the levels—pathways, biological mechanisms, and radiation emitted. They are a much better rule of thumb than the A, B, C classification scheme of 10 CFR Part 61, or the "ten times the half-life" approach, which doesn't account for concentrations.

Waste containing high concentrations of long-lived radionuclides, in general, class C and greater waste under 10 CFR 61, should be treated as high-level waste. Such waste could be stored deep underground, in a high-level waste repository, or stored at the reactor. This is the 10,000 + -year category. This waste category represents a small volume consisting almost entirely of activation products, that is, in general, stainless steel which has been heavily bombarded with neutrons. As seen in Chapter 1, these wastes will arise

Figure 4-2

"Low-Level" Waste

☐ waste hazardous for about 100 years
▨ waste hazardous for about 300 years
■ waste hazardous for over 10,000 years

Waste should be segregated into waste streams according to longevity. If "low-level" waste were segregated according to half-lives, a decreasing volume of waste would need to be maintained with time. After 300 years, only 0.5% of the original waste volume would still be hazardous. On the other hand, if all "low-level" waste were mixed, the entire volume would have to be maintained for greater than 10,000 years.

in great radioactive quantities when power reactors are decommissioned.

A grey area is waste in the 300-year category, composed primarily of waste from utilities. It includes waste from cleaning reactor coolant, concentrated liquids, for example, but also other wastes from a handful of radioisotope suppliers. These wastes should be stored retrievably in more substantial, shielded structures, with leachate collection systems. This waste represents less than 5 percent of the "low-level" waste volume.

Storage/Disposal

The nuclear industry, like any business, attempts to maximize profits by cutting costs. The cost of managing waste, the unwanted byproduct of nuclear power production and radioisotope use, is seen as one cost to be minimized.

The practice of dumping toxic chemicals or radioactive materials into holes in the ground and covering them with earth was obviously cheap to the generator, but not to the public at large. Within a short period of time, less than 10 years in humid environments, steel barrels and cardboard and wooden boxes degraded. Water entered trench cavities and radioactive materials began moving from landfills. But, by then, title and costs had passed from the waste generators to state governments.

The pendulum is now swinging from "out of sight, out of mind," to "always in sight, always in mind," that is, towards above ground storage with title and costs remaining the responsibility of the waste generators. The storage/disposal technologies reviewed here all include engineered structures and solidified waste forms within improved containers, and are vast improvements over old-time shallow landfill. The engineered systems examined here include above ground, below ground and deep underground storage facilities, all discussed in relation to specific "low-level" waste streams.

All waste management systems must be designed to prevent water in-migration and the leakage of radioactive waste, and to minimize occupational exposures. Landfills in humid environments have not been able to accomplish these tasks for periods of just 10 years, let alone for the hundreds or thousands of years required to isolate radioactive waste. Since much of "low-level" waste will be hazardous for more than 100 years, that is, past the assumed institutional control period, the institutional relations, contractual agreements and funding mechanisms must endure for long periods of time. Surveillance, monitoring and maintenance must be a continuing exercise to prevent leakage from waste facilities and harm to future generations.

Other than leakage, the primary pathway for radiation to humans is through direct exposure. The concern is that humans will intrude upon a waste facility and perhaps remove irradiated stainless steel, the Pharoahs Tomb Scenario (see page 50). If this radioactive metal were recycled, the radiation dose to the population could be quite high. Thus, each waste storage and disposal system must be evaluated against the possibility of human intrusion.

Above Ground Systems

Storage Building

Several types of above ground storage facilities are presently in use. In its simplest form, an above ground storage facility is an enclosed warehouse. Ontario Hydro (Carter,1985) employs a large unshielded building, 50 meters by 30 meters, for handling waste with radiation fields less than 1 rem per hour. The walls and roof are fabricated of pre-stressed concrete. The design employs smoke detection equipment, a carbon dioxide fire extinguishing system, and a forced air ventilation system. The warehouse also has an internal drainage system, so that any leakage is retained.

Many utilities in the U.S., such as the Virginia Electric Company, have built corrugated steel lag storage buildings. These hold "low-level" waste until sufficient quantities are accumulated to make up a truck shipment to one of the three operating landfills. Waste is either packed into free-standing, self-stacking metal containers, stacked to a height of 6.25 meters, or placed on metal shelves (see Figure 4-3).

With the cost of waste disposal increasing, many institutional generators, such as Dartmouth College (RWC,1985b), are holding radioactive waste until the short-lived radionuclides have decayed. The cut-off is usually a 60-day half-life (iodine-125), or about 2.5 years until decay to non-hazardous levels.

Waste streams with high concentrations of tritium, carbon-14 and other radionuclides are solidified and stored until sufficient volumes are available for a shipment, or until the waste broker arrives. Unfortunately, waste streams with carbon-14 or tritium activity levels lower than 0.05 microcuries per gram are now being disposed of down the drain or in municipal landfills. This practice leads to an increase in general background radiation levels and a build-up of carbon-14 in the environment.

Storage Vaults

More substantial structures, storage vaults, are designed to hold higher activity waste. The Tennessee Valley Authority put four storage vaults into service in 1982 at its Sequoyah Nuclear Plant (NRC,1982c). Sequoyah's storage vault is essentially a large concrete box with outside dimensions of 192 feet by 34 feet and height of 20 feet. The box opens from the top and is serviced by a gantry crane (see Figure 4-4). Separate vaults exist for resins and dry active waste.

TVA built the four vaults in September 1982 in anticipation of a waste disposal crisis. A total of 18 were on the drawing boards. The vaults holding the resins have an epoxy-type, decontaminable coating on their 42-inch thick walls. The trash vault has 2-foot thick walls. The floor slab is 40 inches thick, and the top cap is 2 feet thick (see Figure 4-4).

To minimize occupational exposures, the vaults are loaded remotely, using a closed circuit TV system. As shown in Figure 4-4, a truck carrying a resin liner within a transport cask parks on one side of the vault. A gantry crane operator, using a 15-ton hoist, lifts the resin liner from the cask, operating the crane from controls on the other side of the vault. The gantry crane has two cross beams, the front beam holding a 15-ton hoist, and the rear beam holding two 30-ton hoists to lift the storage vault cap. The diesel-powered crane straddles the vault, riding on concrete runways on both sides of the vault.

The vaults have an internal liquid collection system. Any liquids released from containers are routed to drainage and sampling valves. As long as the plant operates, these liquids can be transferred to the Sequoyah Nuclear Plant for processing by the radwaste system. While the TVA system is designed to handle dry active waste and resins, it obviously could be used to store other radioactive waste such as irradiated components.

The design lifetime of the storage vault is 40 years, the same as the nuclear plant's, but with a maintenance crew, the real lifetime of the facility should be considerably longer, as long as the lifetime of stainless steel and cement. This assumes, of course, that deterioration of cement and steel due to environmental causes, such as acid rain, is repaired. The design lifetime of concrete can be greatly extended with plastic and elastomeric coatings which shield concrete from chemical attack. Coatings, such as epoxy, polyurethane, asphalt, coal tar and chlorinated rubber, have been suggested (NRC,1986d). Brookhaven National Laboratory has concluded that, with appropriate coatings, concrete is a satisfactory structural material.

> Review of the literature on concrete, including results of testing and modelling, has led to the conclusion that its durability is great enough that it can perform satisfactorily as a structural material in alternative low-level waste disposal methods. There are a number of examples of ancient

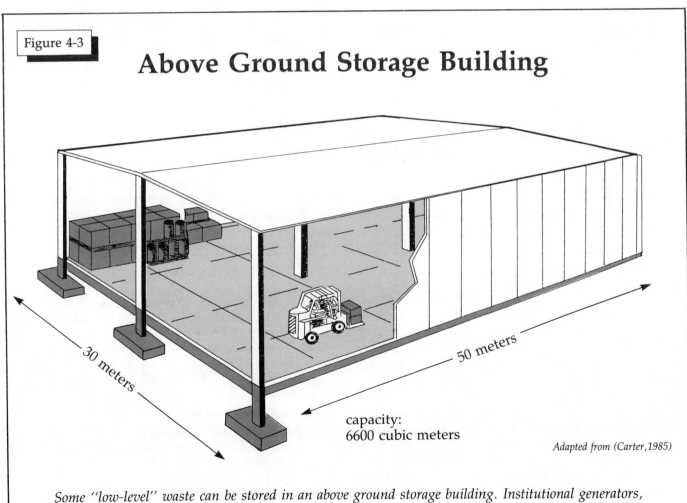

Figure 4-3

Above Ground Storage Building

30 meters · 50 meters

capacity: 6600 cubic meters

Adapted from (Carter, 1985)

Some "low-level" waste can be stored in an above ground storage building. Institutional generators, such as Dartmouth College, and utilities, such as Ontario Hydro, store waste for decay.

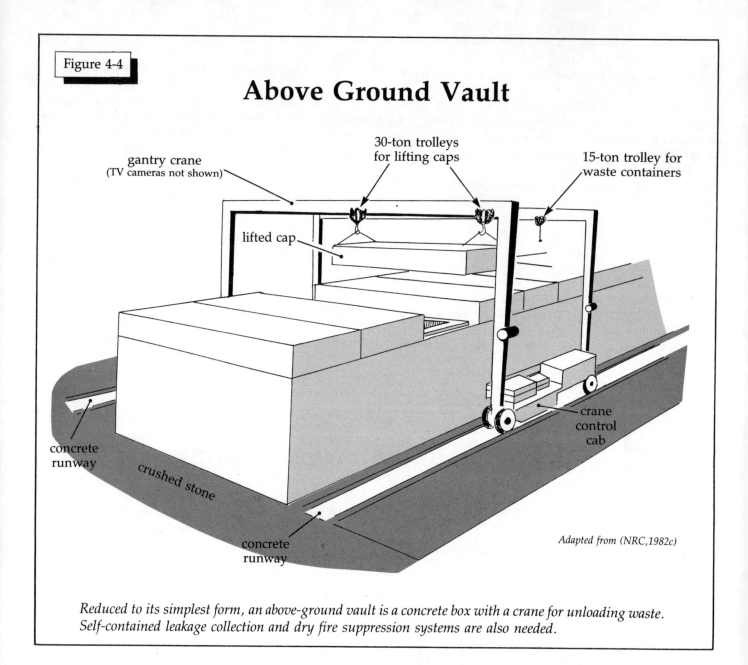

Reduced to its simplest form, an above-ground vault is a concrete box with a crane for unloading waste. Self-contained leakage collection and dry fire suppression systems are also needed.

concrete which have lasted thousands of years and are still in good condition (NRC,1986d).

While the TVA storage vault is essentially a big concrete box, useful for other than reactor waste, other types of storage vaults are more specialized for nuclear operations. For example, the Quadricell is specifically designed to accommodate ion exchange resins and irradiated reactor components (see Figure 4-5). Each Quadricell is divided into four cells, into which fit cylindrical concrete vessels holding resin liners. Like row houses, fifteen 24 cubic meter capacity Quadricells are placed in a row at Ontario Hydro's Bruce Station. As seen in Figure 4-5, the Quadricells have an internal and external leachate collection system. The design lifetime is 50 years.

While the Quadricell is designed for a specific purpose, it has general features common to all above ground storage facilities, namely, an above ground storage box or vault, internal and external leachate collection systems, remote handling systems for high activity waste, and forced air ventilation and carbon dioxide fire extinguishing systems.

Since these structures are extremely sturdy and sit above the ground, geological siting conditions are not as critical as for below ground systems. Above ground structures can be sited almost anywhere, and nearer to where waste is generated. From the nuclear industry's perspective, the major drawback of above ground facilities is their higher cost and the general perception that storage is not disposal.

In the U.S., Westinghouse's Surepak system, hexagonally-shaped concrete modules (see Figure 4-6), can also be used for above ground storage. Rather than bringing waste to the storage system, the Surepak is loaded in a central processing building and brought to the concrete pad where

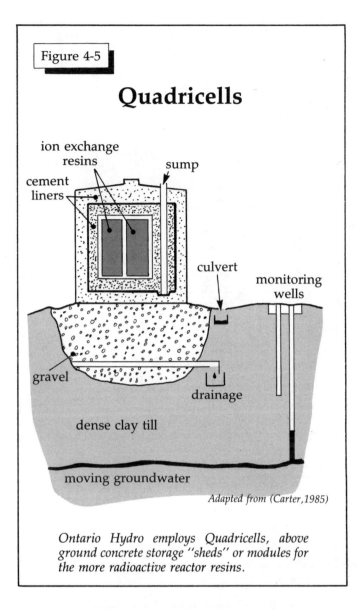

Ontario Hydro employs Quadricells, above ground concrete storage "sheds" or modules for the more radioactive reactor resins.

long-term storage/disposal takes place. In the processing building, dry active waste can be supercompacted into drums that are inserted into the storage module. Resins or high activity components can also be placed in a module, with the interstices filled with cement. The entire module is then transferred to the waste disposal/storage area. The hexagonal shape allows for a tight storage configuration. If placed on an above ground storage pad, a separate and external leachate collection system has to be constructed. As Figure 4-6 shows, Westinghouse has suggested that the Surepak system be used for below ground storage. Since water can pass through concrete, this would not be as safe as using it for above ground storage.

The basic difference between a vault and module storage system is that vaults can be designed with internal leachate collection systems, whereas the smaller storage modules use a common external leachate collection system. Leaks can therefore be detected in a vault before leachate moves outside. In addition, there is some doubt whether individual leaking containers within a Surepak can be recovered and repaired. Cement grout between individual containers within a Surepak make it virtually impossible to recover containers without destroying them.

A hydrogen gas explosion (Siskind,1986) is a risk in a poorly designed above ground facility. Alpha and beta radiation of water breaks H_2O into its constituents, hydrogen and oxygen gas, which can explosively recombine. Water is abundant in radioactive landfills, but is also abundant in certain waste forms, whether stored above or below ground. While NRC regulations prohibit waste forms which contain more than 1 percent free-standing liquid, in fact, ion exchange resins can contain up to 50 percent liquid in interstitial spaces.

Two conditions would increase the probability of explosions in waste facilities: excessive radiation loading of ion exchange resins (the range extends from 0.1 to 30 curies per cubic foot, a factor of 300), and a confined, that is, non-ventilated space.

If an explosion were to occur, radioactive materials would be spread throughout the interior of a storage facility. It is doubtful that the explosive force would be sufficient to rupture the container and the above or below ground storage facility itself, but occupational exposures would likely increase in the cleanup. Other waste containers might be damaged as well. It is unlikely that much radioactivity would be released to the environment, and population exposures would be minimal. Unfortunately, the NRC has not carried out calculations to evaluate this hazard, which could be mitigated with proper ventilation.

A quieter, but more deadly, accident is the inadvertent recycling of radioactive stainless steel into everyday products, such as tables, chairs or the million-and-one other uses of stainless steel. An incident like this occurred in Juarez, Mexico in 1984, when radioactive cobalt was recycled into steel table legs. It was only detected because some tables being brought *into* Los Alamos National Laboratory registered on radiation detectors.

Reactor internals remain radioactive for long periods of time, tens of thousands of years. If reactor internals were recycled, a direct gamma and X-ray dose would be spread over millions of people for tens of thousands of years. It is beyond the scope of this book to estimate the number of health effects, but the NRC should clearly take it upon itself to do the calculations.

Advantages and Disadvantages

An above ground storage system, with multiple engineered barriers and leachate collection systems, provides far greater protection against radioactive leakage than the old "kick and roll" method of waste disposal in which drums were rolled off the end of a truck and covered with a bulldozer. Since steel drums can deteriorate over time, above ground storage is superior no matter what drum stacking method is used.

All the problems that can occur to waste in extended

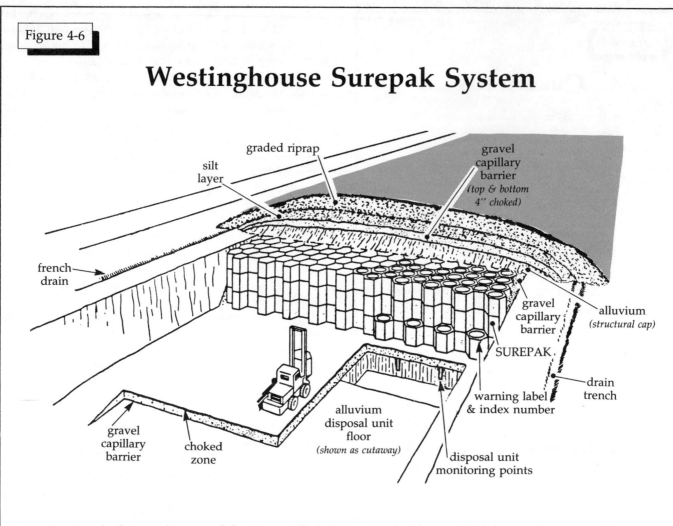

Figure 4-6

Westinghouse Surepak System

The Westinghouse storage module system called the SUREPACK can be used to store radioactive waste above or below ground. The concrete modules are loaded in the waste preparation facility and moved to the storage pad.

storage (NRC,1985b) occur in a landfill. Some problems are magnified in extended storage, but some are magnified in landfills. The major difference is that leakage from an above ground storage facility can be easily detected and repaired. Not so at a landfill.

If a landfill were leaking, most likely the site operator would not exhume the waste contents to keep them from spreading. An environmental assessment would presumably show that the occupational exposures or the economic costs of safe exhumation would be too high. That is, just when exhumation would be needed to stem downstream contamination, state and federal officials would be reluctant to instigate the process.

Other than generally futile efforts to stem water in-migration to trenches, waste will remain buried in a leaking landfill, and it will continue contaminating the environment. At West Valley, burial trenches continue to fill with water, and must be pumped to prevent trench covers from breaking through. This pumped water is then treated for specific radionuclides, but some radionuclides such as tritium, are released quantitatively, a process that can best be called "controlled leakage." Thus, a decision has been made to allow downstream contamination at West Valley, and this is the norm at Maxey Flats, Kentucky and Sheffield, Illinois, as well.

In a landfill, waste drums corrode, exposing the contents to groundwater and the surrounding earth. Radioactivity can migrate, necessitating expensive remedial action. On the other hand, if a corroding container is in a warehouse, the waste can be repackaged. A roof on a storage facility more effectively halts water infiltration than a trench cap, and is more easily repaired. Leaks can be easily detected and repaired. And subsidence problems are less, too. While the problems of gas generation, methane and tritiated vapor, are certainly serious, landfills are vented, and storage buildings can be vented as well.

While the NRC has often stated that it is supportive of greater confinement systems, it has never been enamored with alternatives to landfills. The Commission has published a report on problems with extended storage (NRC,1985b). To keep a balance on the issue, the NRC should perhaps publish a companion piece entitled, "Landfills: Potential Problem Areas."

The Commission has raised a series of objections to extended storage which will clearly protract licensing proceedings and make it very difficult to license anything other than the simplest variant of a landfill. The arguments raised are essentially red herrings which obscure the bottom line objection—the systems are more expensive to the generators and will make clear to all that radioactive waste requires eternal vigilance. Of course, the record now clearly shows that landfills, in the end, are very expensive, and also require eternal vigilance. Nevertheless, the Nuclear Regulatory Commission continues to make the misleading distinction between storage and disposal: storage is considered by the NRC to be temporary, whereas disposal is considered final.

The NRC has raised questions about the behavior of the waste form and container during storage, and the effects of extended storage on the properties of the container *after storage*. Specifically, the NRC (NRC,1985b) is concerned that radiolytic gas generation and radiation-enhanced degradation of resins will lead to loss of radionuclides. The NRC is also concerned that during long periods of storage there could be a weakening of steel containers due to corrosion, and radiation-enhanced embrittlement of high density polyethylene.

If reactor resins are subjected to high radiation doses, gases, predominantly hydrogen gas, build up and pose a fire or explosive risk. Radioactive carbon dioxide or methane is also released. Biowaste degrades during storage, generating gas in the process. And paper, if not heavily irradiated, composts in the presence of moisture. None of this is different from what occurs in a landfill, except that an enclosed space, such as an unventilated vault, may explode or catch fire. Thus, vaults and buildings must be ventilated.

Radiation bombardment of resins may lead to acidic solutions within containers, which is very destructive of cement. In a humid environment, carbon steel containers will rapidly corrode, losing their structural integrity. High density polyethylene used in high integrity containers can be embrittled by radiation. Packages that are leaking or weakened, and which require subsequent movement, will obviously need to be repaired. Depending on the remote systems developed, maintenance of containers may cause additional occupational exposures.

Of course, all of these phenomena occur below the ground as well as above. The important difference, as noted above, is that leakage can be more easily and rapidly detected and halted above ground. When containers degrade in a landfill, leakage is not detected until the leachate reaches the monitoring well, assuming there is one, and assuming the leachate goes in that direction. Containers that degrade in landfills will be difficult to exhume.

Other problems, however, are exacerbated in above ground storage. For example, if the building is not maintained above freezing temperatures, freezing and thawing cycles may affect compressive strength of cement. Of course, countless buildings are built of concrete and remain standing during freezing weather. As previously noted, gas generated in an enclosed space can catch fire or explode. Clearly above ground buildings must be ventilated.

The NRC fears that if leachate escapes an above ground structure, it can travel rapidly on the surface to streams since the additional barrier of the earth buffer is not present. Just this problem occurred at the West Valley landfill when leachate overflowed the trenches and moved along the earth's surface. This objection is really another red herring, though, since systems can be and have been designed for internal and external leachate collection.

Neglected by the NRC is the fact that burial in a landfill creates enormous pressures on high density polyethylene containers used as high integrity containers. According to Jur and Poplin, "structural properties of unreinforced plastics, specifically, polyethylene, are simply insufficient to handle the high external loads associated with burial (Jur,1986)." Radiation-induced embrittlement and external ground pressures are likely to cause the so-called high integrity containers to fail.

With above ground storage, the structures themselves provide the structural stability. Package strength is not so critical, since tons of earth do not have to be supported by the waste package, as is the case with below ground storage. Above ground structures reduce the dependence on site characteristics, and may therefore be located at a wide variety of sites.

The NRC has expressed concern that above ground structures may cause higher occupational exposures. Since the earth serves as a radiation shield, the NRC reasons, occupational exposures are lower in below ground "disposal," whereas workers can be directly exposed by radioactivity from waste in above ground structures. The argument would have merit if waste were truly "disposed of" below ground, but such has not been the experience. Occupational exposures can be very high when a leaking landfill must be exhumed (NRC, 1980). Occupational exposures and dollar costs are trade-offs. As shown by the TVA above ground vault system, remote tv cameras and equipment can reduce occupational exposures to minimal levels. Similarly, degraded containers could be repaired in shielded hot cells using remote manipulators. Thus, with appropriate technology, above ground vaults do not necessarily imply higher occupational exposures.

Human Intrusion

The primary objection to above ground storage is the possibility of human intrusion—not inadvertent intrusion, obviously, but intentional intrusion. Some waste streams, such as irradiated components, are long-lived and will remain hazardous for tens of thousands of years. Since above ground structures are so visible, surveillance, monitoring and repair must continue for the duration. Funds will have to be set aside and trusteeships established to safeguard waste for the time periods required. Left untended, waste mausolea will eventually be opened, and the valuables, stainless steel, removed. The fear is that stainless steel will be recycled, similar to the Juarez incident. By properly separating long and short-lived materials, a reduced volume of waste will have to be monitored in the future. Needless to say, those waste generators who contribute to the long-term waste burden should pay increased costs. Perhaps title should remain with the generators as well.

Below Ground

Below ground storage and disposal systems range from earth-mounded concrete bunkers, that is, structures initially at the earth's surface and mounded over with earth, to below ground vaults, augered holes, underground mines, and deep underground repositories being investigated for high-level waste. Below ground vaults can be designed identically to above ground vaults. Earth is stripped, the vault is constructed, loaded and closed, and following the closure of a series of vaults, earth is mounded over the whole facility.

Earth-Mounded Bunkers

The French disposal system at La Manche is a variant of an underground bunker (see Figure 4-7). A deep pit is excavated and a concrete floor and leachate collection system laid. Class B and C "low-level" waste is placed in an underground concrete box. Concrete is poured in and the packages and concrete form a concrete monolith. These are stacked in pairs with a 2-meter void space. Class C and greater "low-level" waste, with high surface radiation levels, is lowered by crane into the void. When the void is full, concrete is poured in, joining the two monoliths. On top of the concrete monolith, class A drums are stacked, with fill placed between the drums. A layer of clay is then covered

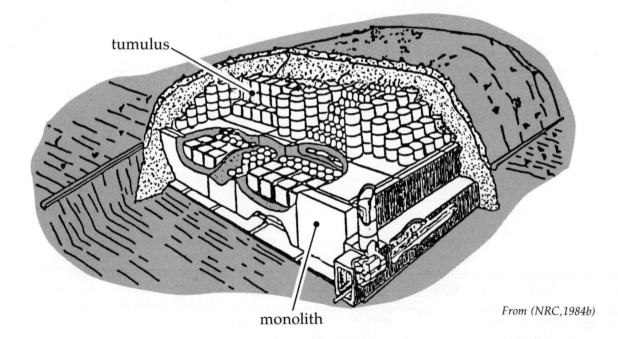

Figure 4-7

Earth Mounded Concrete Bunker LLW Disposal Facility

tumulus

monolith

From (NRC, 1984b)

The waste facility in La Manche, France, consists of below ground concrete blocks or monoliths for storing more concentrated and longer-lived radioactive waste, and drums stored above-grade for less concentrated waste.

with topsoil and reseeded. Internal drainage pipes capture excess water.

The layer of drums on top is called the tumulus. With the massive use of concrete, the base of the structure is obviously highly stable. It is expected that the sloping cover of drums above the monolith will degrade as the barrels corrode. Unless the clay cover is repaired, water will enter the monolith pits in increasing amounts. The weight of the concrete monolith should also raise the water table. Thus, the French system will require continual monitoring and maintenance.

Augered Holes

Any discussion of below ground storage of radioactive waste always makes passing reference to augered holes. Several augered hole experiments have taken place at Department of Energy facilities. Augered holes are drilled with a wide (10 foot diameter) auger bit. Some holes are as deep as 120 feet. Waste is stacked within the hole, sometimes lined, sometimes not. The hole is backfilled and capped with concrete. The waste is therefore non-retrievable.

In many regions, a hole 120 feet deep would place wastes within the water table. At the Savannah River Plant, shallower holes are drilled, down to 30 feet, still allowing a distance above the water table (Cook,1984). Waste at SRP is placed in a grouted fiberglass container (see Figure 4-8).

At the NRC-licensed burial ground at West Valley, waste was placed in unlined holes down to a depth of 50 feet. However, the holes at West Valley regularly fill with water, and there has been tritium seepage from the NRC-licensed burial ground as well as plutonium migration.

Augered holes are not much different in design from conventional shallow landfills, and are customarily used to shield higher activity materials. At Ontario Hydro, concrete pipes line an augered hole. Highly loaded ion exchange resins are grouted with cement and placed within this special auger hole, called a tilehole. A drainage system is used to capture leachate.

Mined Cavities

Previously used underground mines have also been considered for "low-level" waste disposal. The types of mines available fit into four categories: coal, metal, limestone and salt.

Most mine environments are quite corrosive. Coal mines have acid runoff, and require ventilation to prevent gas buildup. Metal mines, in the presence of sulfide mineralization, would corrode steel drums and concrete. Coal and metal mines may have fallen roofs and irregular passageways which have followed the ore or coal seams. Because of the corrosive conditions, waste may not be retrievable from coal or metal mines. Limestone mines, on the other hand, are usually dry and have been used for warehousing. Salt mines are also dry (though salt crystals contain trapped moisture) and like limestone, have regular-sized rooms which can accommodate equipment.

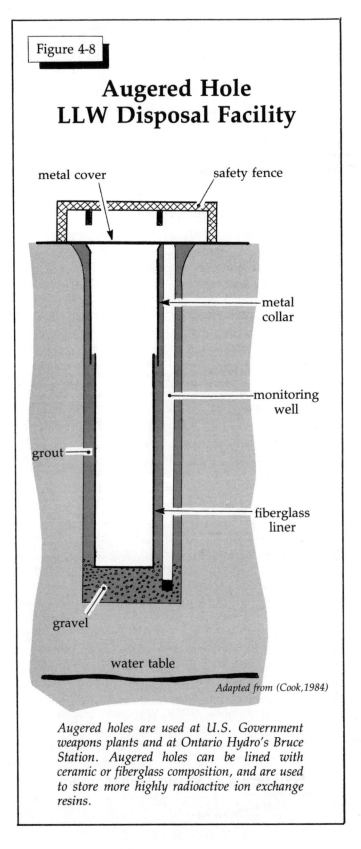

Figure 4-8

Augered Hole LLW Disposal Facility

Adapted from (Cook,1984)

Augered holes are used at U.S. Government weapons plants and at Ontario Hydro's Bruce Station. Augered holes can be lined with ceramic or fiberglass composition, and are used to store more highly radioactive ion exchange resins.

Of course, previously used mines have been developed for resource exploitation and not "low-level" waste storage. Therefore, such mines would have to be completely characterized for this different use. Needless to say, hydrologic

LIVING WITHOUT LANDFILLS 71

conditions would be a vital consideration. New mines could also be constructed for "low-level" waste storage. Sweden is carving out an underground mine in granite for just this purpose.

An important factor in using any mine is future resource exploitation. If a mine has been exploited for salt, potash, metals, coal, limestone or other minerals, it is possible that this could occur again once the institutional control period is past. Inadvertent intrusion into a "low-level" waste disposal area might allow intruders to mine out irradiated stainless steel.

Deep Underground Repository

The Department of Energy is presently engaged in a search for a deep underground high-level waste repository in the states of Washington, Nevada and Texas. A repository would be located 1000 to 3000 feet below the earth's surface. The Department is also mining out an underground area for plutonium-contaminated (transuranic) waste near Carlsbad, New Mexico.

The total volume of all "low-level" waste streams is much greater than that from high-level waste. Thus, disposal of all "low-level" waste in a high-level waste repository would take up massive amounts of space, effectively interfering with the high-level waste program. On the other hand, though, irradiated non-fuel reactor components are small in volume and intensely radioactive, and should be considered high-level waste. The class C and greater waste from decommissioning a reactor is equal to about one-third of the volume of 40 years' worth of irradiated fuel. Thus, it is possible to include this "low-level" waste in a high-level waste repository.

The siting criteria for a high-level waste repository will be more rigorous than for a "low-level" waste facility. The prime concern here is that the present three sites may not be technically feasible or available. Assuming one is licensed, according to the U.S. Department of Energy, the earliest year of operation would be 2003, though most observers expect a much later date.

If irradiated components and ion-exchange resins were included as high-level waste, they would have to be stored until a repository were available. Keeping them at reactors is the most sensible option. Most power reactors will not be decommissioned until about 30 years after reactor close-down, or about 2030, thus the timing is right.

Economics of Waste Management

"There are lies, damn lies, and statistics."
Mark Twain

Though this report touches lightly on economic issues, since the focus here is on safety rather than dollars, it is important to devote some attention to the relative costs of waste management systems. With a finite amount of money to spend, we want to buy real safety.

The technology of waste disposal is not mature. Though humans have been digging holes and burying waste in the ground for thousands of years, the technology for containing toxic materials is a proven failure. Though cost estimates for a waste facility have a certain reality to them, since, to many people numbers suggest accuracy, they are not the same as construction estimates for a bridge or building. Because the technology is immature, the uncertainties can be rather large. Depending on the assumptions, real and imagined, that enter an analysis of waste disposal/storage costs, almost any conclusion can be drawn about the propriety of above ground vs. conventional landfills, or "going-it-alone" vs. joining a multi-state compact. Thus, an element of caution is warranted. In this section, the basic assumptions which underlie available cost estimates, and the basic uncertainties which cast doubt on those estimates, are laid out.

As an example of the inscrutable logic behind economic calculations, consider the following: Is it true that the larger the waste facility, the lower the unit disposal costs? Generally yes, because fixed costs are amortized over more units. But, a large facility implies a multi-state facility, probably a compact, with less democratic siting procedures and much greater citizen resistance. Thus, while a larger disposal facility should mean lower disposal costs, in fact, the pre-operational costs to site and license a facility, and the incentives for a host state and community ("bribes" or "off-sets") can be high. What if the political opposition is too great and a multi-state facility cannot be sited? Realistic pre-operational costs must be included, but the exact number may be hard to come by. Construction costs, on the other hand, are probably the easiest to calculate.

Or, how does one contrast the costs of an above ground vault with a landfill? For a waste facility of the size needed for the Midwest Compact, an above ground vault is about twice as expensive per unit volume. But then, how does one factor in the cost of leakage from shallow land burial?

Past experience at Maxey Flats, Kentucky, and at thousands of toxic chemical landfills, shows that landfills leak. At Maxey Flats, the cost to "secure" a leaking landfill is 10 to 100 times the original disposal cost. Will past history repeat itself? How does one factor in the near certainty of radioactive leakage?

A balance must be struck between short-term internal costs of above ground vaults vs. long-term external costs to the population. One set of costs, to construct an above ground storage facility, is precise and easy to estimate. The other, the health effects to all future populations for the hazardous life of the materials, is vague, indeterminate, difficult to estimate, but likely to occur. It is simple to determine the explicit short-term costs of construction, operation and closure of a waste facility, and to amortize this cost over the number of units, the number of cubic feet. But when waste remains radioactive and hazardous for tens of thousands of years, it is very difficult to estimate the long-term health costs.

Even assuming the total costs are determined, then

who pays what? How are these costs divided among waste generators? Charges are presently based on two factors, volume and curie content. Since land and operational costs are a cost factor, a charge based on volume makes sense. Further, since materials with greater activity have a higher surface dose and are more difficult to handle, requiring a large crane, for example, it makes sense to have a curie charge. The Barnwell facility charges more for containers with higher surface readings.

Yet another cost is the lifetime of the waste facility, which is correlated with the hazardous life of waste itself. Waste generators should pay more for waste that is hazardous for 10,000 years, and pay less for waste that is hazardous for 100 years. A "longevity" charge would encourage waste generators to segregate waste by half-life and would allow a waste facility to maintain separate facilities for long and short-lived materials. This practice would mean that only a small waste volume would have to be watched for long periods of time, as shown in Figure 4-2.

To estimate the cost of a waste facility, the costs can be divided into the following categories: pre-operational, capital, operating, closure and institutional control period.

Table 4-1

Unit Disposal Cost Estimates
(dollars/cubic foot)

	TRW/RAE	EPRI	Dames & Moore	Westinghouse	DOE	RAE	Sargent & Lundy
year of dollar value	1980	1982	1985	1990	1984	1985	1985
annual volume (thousands ft³)	1800	200	1600	150	18	940	210
shallow land burial	6.78	10.00	12.20		23.00		30.80
improved shallow land burial	7.28				40.00		
above ground vaults			23.40				37.50
below ground vaults			25.40				38.50
modular concrete canister disposal			18.90	33.00			40.80
earth mounded concrete bunkers tumulus monolith						8.50 13.90	
mined cavities	32.00		37.50		142.00		
augered holes					48.00		

Source: (Midwest, 1986a)

Pre-operational costs include site characterization, licensing and administrative costs. An important consideration is whether the waste facility will be located at the site of a nuclear reactor or on virgin land. Since the reactor itself is the overwhelming contributor to long-lived waste, the costs of a waste facility are obviously greatly reduced if the reactor itself is the container for 10,000 + -year waste. Can a state require a private utility to store others' waste? Is a state legally required to accept a utility's 10,000 + -year waste?

Capital costs include land, access roads, waste facility construction costs and equipment. Pre-operational costs may be three times capital costs (Maine,1985), and therefore the costs of a facility may not be greatly affected by size.

Operating costs include labor, monitoring and regulatory costs. If the waste facility is a landfill, costs of trench excavation are considered operational. If a facility is an above ground storage system, costs for above ground vaults may be considered a capital cost. Operational costs must also factor in whether a facility operates year-round. A proposed Maine facility, a relatively small facility accepting between 10,000 and 25,000 cubic feet per year, is projected to operate only two months out of the year, except for a five-year period when the Maine Yankee nuclear reactor is decommissioned.

Closure costs include building removal, labor, monitoring, site maintenance and licensing. During the institutional control period, the cost categories are labor, monitoring, administration and insurance. Direct labor costs during this period assume occasional site inspections by security personnel.

None of the above costs includes major repair of an eroding or leaking facility, or exhumation. Assuming a substitute site could be found to which the contents of the exhumed waste facility could be moved, the NRC estimates the cost of relocating a waste facility to be in excess of $1 billion (in 1978 dollars). All calculations usually assume the institutional control period is 100 years, when, in fact, radioactive waste will remain hazardous for considerably longer periods of time.

For the state of Maine, which generates between 10,000 and 25,000 cubic feet per year, the costs of the above categories are estimated to be as follows:

pre-operational	$4–$5 million
capital	$1.6 million
operational	$800,000/year
closure (year 1)	$1 million
closure (year 2+)	$300,000
institutional control period	$50,000/year

Assuming a 5 percent real interest rate, and a 3 percent rate of return on a sinking fund and other money accounts, the projected costs for shallow land burial are $129 per cubic foot in 1985 dollars (includes decommissioning waste). In contrast, estimated costs for the Southeast Compact (Dames,1985), which generates about 100 times as much waste volume as Maine, are $12.20 per cubic foot (1985 dollars). If Maine were to join with New Hampshire and Vermont, the disposal costs would drop to $44 per cubic foot.

Clearly, waste generators have an economic incentive to force states into compacts. Of course, in what seems to be a pattern in other regions as well, Maine residents may not be thrilled with the prospect of taking waste from other states. The difference in cost to Maine residents between "going-it-alone" and joining with Vermont and New Hampshire amounts to $0.71 per Maine resident per year, a relatively insignificant amount.

In determining the cost of a waste facility, an important consideration is whether the facility will be operated by a state authority (with private contracting assistance), or entirely by the private sector. Assuming a 20 percent rate of return and no state income tax, the state of Maine estimates $304 per cubic foot for the private sector option vs. $129 per cubic foot for a state authority.

To contrast shallow land burial with other technologies, cost estimates collected by the Midwest Compact (Midwest,1986b) are compared in Table 4-1. For a large facility, the proposed Southeast facility in North Carolina, Dames and Moore estimates that above ground vaults will cost twice as much as shallow land burial. But, for a facility that accepts only 210,000 cubic feet a year, as is projected for the Midwest, the difference between improved shallow land burial and above ground storage vaults is very small. Though Maine did not make this comparison, for a single state with modest waste volume, the difference between improved shallow land burial and above ground vaults is minor. As seen at the bottom of Table 4-1, for the Midwest Compact, above and below ground facilities are comparably priced. These costs do not include possible leakage from, or repair of, a waste facility.

Table 4-1 also shows that the difference in costs between a state "going-it-alone" or joining a compact is not great. For example, for above ground vaults, Dames & Moore estimates $23.40 per cubic foot for an above ground vault, assuming 1.6 million cubic feet per year are accepted, whereas Rodgers and Associates estimates $37.50 per cubic foot for a facility which receives only 210,000 cubic feet per year. The cost per resident is minimal, close to $0.10 per resident per year. An economic analysis by the Clean Water Fund of North Carolina shows the difference between North Carolina "going-it-alone" or receiving wastes from the entire Southeast is only about $0.61 per North Carolina resident per year (Stegman,1987).

CHAPTER 5

Conclusions and Recommendations

CHAPTER 5

Conclusions and Recommendations

The siting of new "low-level" waste facilities will be exceedingly difficult. Compared to conventional landfills, new and genuinely improved facilities will be proposed. Industry and state experts will "prove" that the harm to the public, from now to eternity, will be minimal, especially compared to everyday risks like driving a car or walking across a street. Citizens and local governments will distrust the so-called experts, and will see the state and the nuclear industry as not disinterested parties to the dispute. Environmental and community activists will resist. The state and waste generators will say the public is irrational. Charge will follow countercharge as the clock winds down to January 1993. Considering the many steps needed to license and operate a new "low-level" waste facility, and the resistance that will be encountered, the timing is very short. The final outcome of the struggle cannot be predicted.

No policy or method for dealing with nuclear waste can be a final answer, and anyone who makes such a claim is either deceptive or foolish. It is simply not possible to guarantee that a plan will work for thousands of years. Thus, the first step in addressing this issue is to recognize that any option for handling radioactive waste will require perpetual scrutiny, eternal vigilance.

That being said, though, it is still quite clear that not all management alternatives are created equal—that is, some are worse than others. Our goal must be to seek the most sensible, environmentally-sound, safety-first approach. In the following section, we provide guidelines toward this end.

The Search for New Dumpsites Should Be Halted

As shown in Chapter 1, based entirely on NRC data, and including decommissioned power reactors, the nuclear power industry generates 99 percent of the "low-level" waste radioactivity, and an even greater percentage of the long-lived radioactivity. Thus, 99 percent of "low-level" waste radioactivity is generated at 115 reactors, at 72 individual sites. This includes all reactors which have operated before and up to 1987, and all that are projected to be operating by the end of 1987.

The obvious first question is—does it make sense to site 10 to 12 new waste facilities if 99 percent of the total radioactivity and 70 percent of the pre-compacted volume already resides at 72 sites in most states in the U.S.? We think not. We note that industry spokespersons are already acknowledging that only two national sites are needed for "low-level" waste (White, 1987). In fact, no new sites are needed.

The search for additional "low-level" dump sites must be halted. All nuclear power plant "low-level" waste should be held at reactor sites and the tiny (in terms of radioactivity) percentage of "low-level" waste generated by medical, institutional and industrial users should be transported to reactor sites. This additional waste represents 30 percent of the "low-level" waste volume before compaction, and a smaller percentage after supercompaction. Reactors should serve as storage sites until such time as superior locations and methods mentioned below for the more permanent storage of "low-level" waste have been developed.

We recognize that institutional arrangements will be required for transferring waste from industrial and institutional generators to utility waste generators. The answers to the many legal and institutional questions posed by such a policy are just surfacing. Yet, at-reactor storage appears a more logical and environmentally sound interim policy than the divisive efforts to site and construct burial grounds and incinerators of questionable safety.

Manhattan Project II

A massive scientific effort, akin to the Manhattan Project, we call it Manhattan Project II, should immediately be launched to attempt to resolve all aspects of the nuclear waste problem. The Manhattan Project sent the best scientific minds of the time into the desert with a virtually unlimited budget and a research agenda untrammeled by bureaucratic strings. It is clear we need a scientific effort of at least equal magnitude to undo the damage unleashed by the fissioning of the atom.

We are convinced that such a project cannot be under the aegis of those government agencies, such as the U.S. Department of Energy and the U.S. Nuclear Regulatory Commission, which have helped to create the nuclear waste problem and continue to add to it through the promotion of commercial nuclear power and the production of nuclear weapons. An independent new entity will have to oversee Manhattan Project II. All of its proceedings should be carried out openly, and it should be open to a full range of ideas. Its board members should be confirmed by Congress.

Minimize the Production of Radioactive Waste

Without a presently acceptable and safe means of managing waste, the production of radioactive waste is irresponsible and must be minimized. Since nuclear power reactors are the largest waste generators, necessary electricity needs should be satisfied in other ways. Presently operating nuclear reactors should be phased out, with federal monies allocated for the retraining of workers dislocated by the phase-out. Perhaps physicists, engineers, chemists, geologists and other people formerly employed by the nuclear power industry could be utilized in Manhattan Project II. We encourage all groups concerned about radioactive waste dumps to assist groups opposing nuclear power reactor operation.

Waste generated by industry and medical and research institutions requires a separate cost/benefit analysis, but the radioactivity produced is small compared to nuclear power plants. Nevertheless, waste production in those sectors should be minimized as well.

Volume Reduction

After waste minimization, volume reduction and storage for decay should be practiced. To some extent the rapid rise in waste disposal costs has forced waste generators to adopt these practices already. Additional reduction can likely be achieved.

Waste Sorting

For the minimal volume remaining, "low-level" waste should be separated by hazardous life, not by classes A, B and C, but into 100, 300 and 10,000+-year categories (see Figure 4-2). All waste in the 10,000+-year category should be considered high-level waste under our plan. After 100 years, only wastes in the much less voluminous 300 and 10,000+-year categories, or less than 10 percent of the original waste volume, will remain. After 300 years, less than half a percent of the original waste volume will remain to be safeguarded. These materials could be stored in above ground storage vaults, or superior systems and methods developed under Manhattan Project II.

Where should waste with 100 and 300-year hazardous life be located if no new waste facilities are sited? We propose that the waste produced by institutional and industrial waste generators—amounting to 1 percent of the total radioactivity and 30 percent of the pre-compacted volume—be brought to the 72 reactor sites. These reactors should be phased out. This fact must be faced. Reactors are waste dumps that contain 99 percent of the "low-level" waste radioactivity. There is no magic place to take waste—someplace, somewhere else. Less waste must be produced and the remainder must be monitored and maintained. Forever.

What Should be Done with Long-Lived "Low-Level" Waste?

As shown in Chapter 1, long-lived waste, with a hazardous life of over 300 years, consists of irradiated components (metals that have been bombarded with neutrons) and ion exchange resins. Reactor internals, decommissioned reactors and ion exchange resins should immediately be removed from the "low-level" waste stream and reclassified as high-level waste. Keeping these extremely long-lived items in the "low-level" waste category makes about as much sense as putting long-term hardened criminals into the same jail cells as minor offenders. There should be an immediate halt to the current practice of dumping reactor internals and ion exchange resins in "low-level" landfills. The stainless steel in irradiated components, an inviting target for looters, should be treated as high-level waste.

Until the high-level waste issue is resolved, all reactor waste should remain at nuclear reactors for indefinite storage. In some instances, where a reactor is inappropriate for above ground storage because of its location on a floodplain, proximity to earthquake fault zones, or other unacceptable risks, we would recommend that the "low-level" waste materials be transported to a superior reactor site. For the most part, reactors will not be dismantled, and irradiated components discharged, until the year 2030 or later. For various reasons, primarily occupational exposures and the difficulty of cutting up a highly radioactive reactor, it is likely reactors will have a cooldown period of at least 30 years following cessation of operations.

The resolution of the high-level waste problem is beyond the scope of this book. We do lack confidence that the Department of Energy can resolve the high-level waste problem and site a deep underground high-level waste repository. The Department of Energy, the agency which enthusiastically promotes the further proliferation of nuclear materials, cannot, simultaneously, engage in the scientifically rigorous process needed to develop a solution that will make sense now, as well as in the future. Politically-contaminated science has no role to play in what may be the most complex, awesome and intractable socio-technical problem in the history of the human race. Adequate geologic characterization for the time frames involved is, thus far, beyond the scope of 20th century science.

Decision-Making or the Lack of It

Citizens are being asked to involve themselves in decisions about what to do with radioactive garbage. Should it go here versus there? Can citizens speak five minutes or ten? Should there be one hearing or two? Committees are being set up to facilitate "public participation." This entire process must be democratized. There must be substance as well as form.

Unfortunately, thus far, the trend seems to be away from democratization. The regional compacts set up by the Low-Level Radioactive Waste Policy Act of 1980 are an example. The compacts set up a new governing mechanism, compact commissions, with the members appointed by the governors of the involved states. The powers these commissions have vary from region to region; they include the power to select a host state for a dump site, override state siting laws, set financial liability limits, etc.

In other words, the major decisions about a facility which will be hazardous for tens of thousands of years have been taken out of the purview of elected state legislatures and placed in the hands of individuals who are removed from the accountability of the democratic process. Although state governors are the appointers, citizens have a long four-year wait before being able to show through the ballot their opposition to compact decisions. And by the time gubernatorial elections roll around, the key decisions will have already been made. State legislatures have acquiesced without a peep to this federally-imposed erosion of their powers. This process, in effect, contributes to insulating decisions about waste facilities from full public debate.

Regional compact commissions should be made accountable to elected state legislatures or be dissolved. And mechanisms need to be put into place to guarantee that commissions draw their members from a range of viewpoints, including citizen groups.

But the location and technology of a waste facility are only part of the decision. Where are the committees which decide how much and what type of waste production is absolutely essential? What referendum was taken to determine whether a nuclear reactor should have been licensed to operate and to produce waste? The fact is, as things stand now, there is no limit to the amount of waste generation. Local communities are simply being asked to accept however much is being produced, without having any say about whether it should be produced at all. Clearly the process is not democratic.

BRC and High-Level Waste

Two recent trends in radioactive waste management are very disturbing. First, the nuclear industry and radioisotope users, in collaboration with the NRC, are attempting to deregulate waste with low concentrations, allowing those below a certain limit to be placed in municipal landfills. Moreover, the NRC has no regulations against diluting radioactive waste and pouring it down the drain. Which waste streams are BRC, below regulatory concern, and which are not, is on a sliding scale that is expected to rise in proportion to the rise in waste disposal costs. As a consequence, more radioactivity is entering the accessible environment and background radiation levels are increasing.

At the top end, the NRC has instituted a proceeding to redefine high-level waste, the inevitable consequence being that a large amount of highly radioactive waste may be reclassified as "low-level" waste. This particularly includes the high-level waste in tanks at the federal reprocessing facility at Hanford, Washington.

The process must be pushed in the other direction. Citizens should resist industry and NRC efforts to redefine waste from the high-level to the "low-level" category. Long-lived activated metals, such as fuel hardware, in-core instrumentation, reactor internals, sealed sources and plutonium-contaminated equipment, should remain in the greater than class C category and be treated as high-level waste. The volume of class C and greater waste is not a significant addition to the high-level waste generated by the same reactor.

Who Pays?

Radioactive waste, particularly from nuclear power reactors, is a long-term responsibility. Exactly how institutional arrangements and funding mechanisms will be set up to accomplish this task is not clear. But the general principle that those waste generators who account for the costs should pay their share is generally acknowledged. Waste disposal charges are currently based on volume and radioactivity emanating from the container. We recommend that an additional charge be levied, based on the hazardous life of the radioactive materials. Clearly, the contributors of long-lived materials should pay more. Perhaps the generators of this waste should retain title as well, in case problems and additional costs arise.

Eternal Vigilance

The bottom line of our waste management plan is eternal vigilance. Waste remains in sight and in mind. As waste containers and storage vaults degrade, future generations will need to retrieve, repair and replace them. Waste must be stored in ways accessible to future generations. We can no longer produce waste, place it in the ground and hope that the earth stands still. Unless Manhattan Project II and spectacular advances in technology in the 21st Century provide far better containment than we now have, each generation from now on will have to deal with repackaging some of the wastes our generation is leaving for them.

Decisions made over the past 40 years—to use nuclear power plants and build nuclear weapons—have condemned all generations henceforth to eternally safeguard nuclear wastes. These wastes can never be completely out of mind. There can be no Hollywood ending to this story. To pretend otherwise is to invite tragedy.

But once we have recognized the awesome nature of this problem, and after we reject shortcut "solutions," we can take the steps needed to safeguard health and the environment, now and in the future. This book is meant to contribute to beginning that process.

Appendices

GLOSSARY

Accelerator A device for increasing the kinetic energy of charged particles (for example, electrons and protons) through the application of electrical and/or magnetic forces.

Activation Products Radioisotopes formed through bombardment with neutrons or other particles; nuclides such as tritium, carbon-14, cobalt-60 and nickel-59 and nickel-63 are activation products. Transuranic nuclides such as plutonium-240 are also included by strict definition.

Alpha Particle A positively charged particle, identical to the nucleus of a helium-4 atom, consisting of two protons and two neutrons, emitted by uranium and heavier radionuclides. Though alpha particles cannot penetrate clothing, they are extremely damaging when taken internally.

Aquifer A subsurface formation or geological unit containing sufficient saturated permeable material to yield significant quantities of water.

Background Radiation Radiation in the environment that includes the sum of the emissions from cosmic radiation, radioactive materials that occur naturally, and those resulting from the nuclear fission process which have been released to the environment a year or more in the past.

Below Regulatory Concern (BRC) Materials contaminated with radioactive substances in sufficient dilution that they are considered to be unworthy of regulation and, thus, can be disposed of in sanitary landfills or down the drain.

Beta Particle An electron emitted in the radioactive decay of certain nuclides, such as strontium-90 and tritium. Beta particles cannot penetrate heavy metal, but can cause skin cancer, and are harmful when taken internally.

Biowaste Biological matter or animal carcasses contaminated with radionuclides, such as tritium and carbon-14.

Boiling-Water Reactor (BWR) A light water reactor that employs a direct cycle; the water coolant that passes through the reactor is converted to high pressure steam that flows directly through the turbine.

Chelating Agents Chemicals that strongly bond metals into chemical ring structures. Chelating agents are commonly used to facilitate decontamination of radioactive surfaces. Examples of chelating agents include EDTA (ethylenedaimine tetraacetic acid), and TBP (Tri-n-butyl phosphate).

Class A, B or C Division of "low-level" waste by Nuclear Regulatory Commission regulations, 10 CFR Part 61, into categories according to concentrations of specific long and short-lived radionuclides.

Curie A measure of the rate of radioactive decay. One curie is a large amount of radioactivity, equal to 37 billion radioactive disintegrations per second.

Decommissioning The process of removing a facility or area from operation and decontaminating and/or disposing of it or placing it in a condition of standby with appropriate controls and safeguards.

EDTA See chelating agents.

Evaporator A device for reducing the volume of aqueous radioactive waste solutions by boiling off excess water.

Filter Sludge Radioactive cake of powdered resins and cellulose fibers formed on a wire mesh or cloth. The resins and fibers remove radioactivity from reactor water.

Fission The splitting of a heavy nucleus into two approximately equal parts (which are nucleii of lighter elements), accompanied by the release of a relatively large amount of energy and generally two or more neutrons.

Fission Product A general term for over 200 different nuclides of over 35 different elements produced as a result of nuclear fission. Most fission products are radioactive.

Fuel Assembly A bundle of fuel rods and their related hardware which have been arranged for efficient operation of a nuclear reactor.

Fuel Rod A tube, about 14 feet long, containing uranium dioxide fuel stacked like poker chips within; part of a fuel assembly in a nuclear reactor.

Gamma Radiation Electromagnetic radiation, similar in form to light, but much more energetic. Gamma rays are very penetrating and require dense materials, such as lead or uranium, for shielding to stop them.

Half-Life Time required for half of a radioactive substance to lose its activity. For example, in 29 years half of a given quantity of strontium-90 will decay away. In another 29 years only a quarter of the given quantity will remain. As a rule of thumb, ten half-lives are required for a substance to decay to safer levels.

Hazardous Life The time required for a radioactive substance to become non-hazardous, defined here as the time for the radioactive concentration to reach 100 times maximum permissible concentration.

High Integrity Container (HIC) "Low-level" waste container projected to hold its shape for 150 years. When "disposed of" in HIC containers, Class B and C wastes do not also have to be stabilized.

Incinerator Device used to burn toxic and radioactive trash. Radioactive byproducts of the combustion process are collected as radioactive ash or released as gases into the environment.

Ion Exchange Resins Bead-like plastic chemical products which separate and purify chemicals and concentrate waste products, such as cesium-137 and cobalt-60, for "disposal."

Irradiation Deliberate exposure of a substance to radioactivity.

Leachate The polluted liquid containing the soluble components of waste which leaks from a landfill.

Maximum Permissible Concentration (MPC) The amount of radioactive material in air, water, or food that would result in a dose of 500 millirems per year to the whole body, or 1,500 millirems per year to a specific organ, as given in Section II, Table B, of the Nuclear Regulatory Commission regulations, 10 CFR Part 20.

Megawatt (MWe) One million watts of electricity.

Megawatt (t) One million watts of thermal power. In passing thermal energy or steam through a turbine, electrical energy is produced.

Millicurie (mCi) One one-thousandth of one curie.

Millirem (mr) One one-thousandth of one rem.

Nanocurie (nCi) One billionth of a curie.

Neutrons A subatomic particle with zero electric charge and a mass nearly that of a hydrogen atom. Low energy neutrons can induce fission in fissionable materials.

Nuclear Reactor A device in which a controlled chain reaction is maintained, either for the purposes of experimentation, production of weapons grade fissionable material, or generation of electricity.

Nucleus The center of an atom, in which protons and neutrons are located. Plural, nucleii.

Picocurie (pCi) One trillionth of a curie.

Pressurized Water Reactor A light water reactor that employs an indirect cycle; the cooling water that passes through the reactor is kept under high pressure to prevent boiling. Water is heated in a secondary loop that produces steam to drive the turbine.

Production Reactor A nuclear reactor designed for turning one nuclide into another, usually natural uranium into plutonium, for the manufacture of nuclear weapons.

Rad (Radiation Absorbed Dose) The basic unit of absorbed dose of ionizing radiation. A dose of one rad means the absorption of 100 ergs of radiation energy per gram of absorbing material.

Radioactivity Spontaneous emission of radiation due to changes in the nucleus of an atom.

Radiography The production of images by X- or gamma rays on a photographic plate.

Radioimmunoassay Laboratory counting technique employing radionuclides used to tag cells.

Radiolytic Relating to chemical decomposition by the action of radiation.

Radioisotope A radioactive isotope. An unstable isotope of an element that decays or disintegrates spontaneously, emitting radiation.

Radionuclide A radioactive species of an atom.

Radiopharmaceuticals Radioactive drugs used for diagnostic or therapeutic purposes.

Rem (Roentgen Equivalent Man) A widely used unit of measure for the dose of ionizing radiation that gives the same biological effect as one roentgen of x-rays. One rem equals approximately one rad for x-ray, gamma, or beta radiation, and one-tenth rad for alpha radiation.

Reprocessing The chemical process by which unfissioned uranium and plutonium are separated from fission products for further use in new fuel fabrication or weapons production.

Sealed Sources High concentrations of radionuclides within a capsule. Cobalt-60 and cesium-137 sources are used to irradiate humans. Plutonium-238 and strontium-90 sources are used to generate heat. Americium-241 sources are used in smoke detectors.

Sorption In chemistry or geochemistry, the general term for the retention of one substance by another by close range or chemical forces. Absorption takes place within the pores of a granular or fibrous material. Adsorption takes place largely at the surface of a material or its particles.

Source Material Uranium or thorium or any ores that contain at least 0.05 percent of uranium or thorium.

Special Nuclear Material (SNM) Plutonium, uranium-233, enriched uranium-235, and any other materials that the NRC determines to be SNM's.

Subsidence Gradual or sudden sinking of ground surface below grade level due to slow decay and compression of material. Subsidence is a common pathway for water to enter waste trenches in landfills.

Supercompactor A hydraulic press generating 1500 to 2200 tons of force; used to greatly reduce the volume of compactible radioactive waste.

Tailings Waste material from processing uranium. Since uranium ore contains less than 1 percent uranium, essentially all of the processed ore is left as tailings near uranium mills.

TBP See chelating agents.

Transuranics Radioactive materials heavier than uranium, such as plutonium, curium, americium and neptunium. All transuranics are man-made elements.

Tumulus A pyramid-shaped mound of stacked drums or boxes containing radioactive waste. The mound is covered with layers of sand, gravel and clay. Rip-rap is used to prevent human intrusion.

Uranium Mill Tailings See tailings.

Waste Stream The term is used in two ways: as effluent which is released directly into the environment, and as a category of waste which is packaged for a waste facility. Numerous activities use or create radioactive materials, and in the process, generate "low-level" waste. Waste generated by each specific, distinct activity is known as a waste stream. A waste stream may be specific to one producer, or to a group of similar waste producers.

Water Table The upper limit of ground saturated with water. The level to which a well would fill with water.

X-Rays Penetrating electromagnetic radiation, the wavelengths of which are shorter than those of visible light. X-rays are generally less energetic than gamma rays.

REVIEW COMMENTS
by the Nuclear Regulatory Commission
AND RESPONSE
by the Radioactive Waste Campaign

The Review Panel for *Living Without Landfills* (see page iv) made numerous helpful comments and recommendations, almost all of which were incorporated, greatly improving the final report. So that our readers may also see another viewpoint and our response, we are printing a short dialogue with the Nuclear Regulatory Commission (NRC,1987c).

NRC Comment: *All data presented should include specific reference to its source. Computations, equations, tables and figures should also have their sources identified. For example on page 4, the statement is made that "100 nuclear reactors produce 99 percent of the radioactivity in low-level waste in the United States." The Conclusion section states that it is "based entirely on NRC data."*

RWC Response: All information comes from, or is based on, publicly available documents, many of which were produced by the Nuclear Regulatory Commission. We have tried to be diligent in referencing our sources, about 100 of which are listed in References. Information that 115 reactors produce 99 percent of the radioactivity in the United States is based on NUREG/CR-4370, *Update*. All licensed reactors, and those projected to be operating by the end of 1987, as discussed in *Update*, were included. Waste from this fixed electrical generating capacity was projected forward 40 years.

NRC Comment: *The discussions of the experience at closed low-level waste disposal sites should reflect two major points:*
1) While releases of radioactivity have occurred, these releases were small and have not endangered public health and safety.
2) The low-level waste management regulation, Chapter 10, Code of Federal Regulations (CFR) Part 61, reflects the experiences at the closed disposal sites and is intended to eliminate these past problems.

RWC Response: 1) We agree that the releases have so far been small, but they have been totally unexpected and have occurred within ten years of burial. Citizens have lost confidence in the regulatory agencies and their staffs, and in the computational models. This "low-level" waste will remain hazardous for hundreds to hundreds of thousands of years. The total number of potential health effects has not been seriously estimated, and may not be small. Inadequate funds have been set aside to monitor, maintain and repair these radioactive waste dumps. Thus, at the expense of future generations, present day costs are being kept low.

2) We do not share the NRC's confidence that the regulations, 10 CFR Part 61, will eliminate past problems. The draft regulations did "reflect" experience at the closed disposal sites, and were "intended" to eliminate past problems. Not so, however, with the greatly weakened regulations that were ultimately adopted.

Specific numerical criteria were replaced with general guidance which will be translated into numerical criteria in the licensing process. This greatly disadvantages citizens who do not have the resources to keep up with highly paid industry consultants. Those numerical criteria which were retained, specific concentrations in class A, B and C waste, were weakened by a factor of 10, *with no new supporting information.* In the NUREG/CR-4370, *Update,* radionuclide concentrations were greatly increased for some reactor waste streams. For example, iodine-129 concentrations in ion exchange resins were increased by a factor of 230, yet no further actions were taken by the NRC. Iodine-129 has a half-life of 17 million years, and because it is water-soluble, is virtually certain to reach humans. Stainless steel reactor internals, with long-lived activation products, will remain an enticing booty for looters, what we call the Pharoah's Tomb Scenario. The NRC "intruder-construction" scenario assumes that intruders will see large distinguishable blocks, cease looting and check the deeds. Experience at West Valley, Beatty and Juarez clearly disputes the NRC's fanciful scenario. Besides, proposed regulations by the Environmental Protection Agency do not allow agencies to claim credit for monitoring past 100 years. In response to NRC comments, we have written a separate chapter (Chapter 3) on the federal response.

NRC Comment: *The term "hazardous life" is used throughout the document and in Table 1-1. What is the Campaign's technical concept of "hazardous life?" Hazardous life should be defined and the Campaign's method of calculating hazardous life should be appropriately referenced.*

RWC Response: A separate section on "hazardous life" has been added to Chapter 1. As is implied by the NRC's comments, defining "hazardous life" is not a simple matter. Each radionuclide, in differing chemical form, behaves differently in the environment and concentrates in different humans organs. Each radionuclide has specific biological uptake and retention, emits unique radioactive emissions, and gives rise to specific health effects. The "10 half-lives" rule of thumb is one indicator of how long radioactive material remains hazardous, but it does not account for the above factors. We have therefore gone back to the regulations used before 10 CFR Part 61, namely, maximum permissible concentration (MPC), defined in Table II, 10 CFR Part 20. We define "hazardous life" as the time for the concentrations of radionuclides in radioactive waste to decay to 100 times MPC, as defined in Table II, 10 CFR Part 20. This is understood to be a rough indicator of the time to regulate, monitor, and maintain "low-level" waste streams.

NRC Comment: *As you have referenced the NRC document entitled, "Update of Part 61 Impacts Methodology" (NRC/CR-4370), one should note that some of the waste streams that are included in the NUREG/CR-4370, your Tables 3-2, 3-3 and 3-4 are greater than class C wastes. Greater than class C wastes, which NRC found generally unacceptable for near-surface disposal, are now the responsibility of the Department of Energy (P.L. 99-420, Section 3(b)(1)(D)) and it is unlikely that States will accept these wastes for near-surface disposal.*

RWC Response: We agree that certain of the waste streams in NUREG/CR-4370 are presently listed as greater than class C "low-level" wastes, and are the responsibility of the Department of Energy. Without an intervention by Congress, the NRC has no intention of reclassifying class C and greater than class C, as high-level wastes. In fact, the process seems to be going in the opposite direction. The NRC has initiated proceedings on the definition of high-level waste in which some waste presently classed as high-level, such as Hanford waste and fuel assembly hardware, may become "low-level" waste. It is still unclear whether greater than class C wastes will be handled by the states or the federal government. From our perspective, which agency takes charge is not as important as what technology is chosen, what short and long-term provisions are taken for protecting the public health and safety, and how effectively the public can participate in the decision-making process. Since 99 percent of the radioactivity in "low-level" waste is generated at reactors, our policy recommendation is that all "low-level" waste should be stored at reactors.

NRC Comment: *Reference should be made to the Congressional mandate regarding below regulatory concern (BRC) wastes to establish standards and procedures for considering petitions to exempt these wastes from regulation–Section 10 of the Low Level Radioactive Waste Policy Act Amendments of 1985 (P.L. 99-240).*

RWC Response: As we now point out in the text, up to 40 percent of the "low-level" waste volume may be deregulated by NRC actions. This is a disgrace and a major abdication of regulatory responsibility.

NRC Comment: *The report's technical discussion on financial assurance does not acknowledge Chapter 10 Code of Federal Regulations Part 61, sections 61.61, 61.62, and 61.63 regarding applicant assurance, funding for disposal site closure and stabilization, institutional controls, and long-term care funds.*

RWC Response: Since the institutional control period is 100 years, we are not greatly reassured about these provisions in Part 61. Further, no licensing proceedings have taken place under 10 CFR Part 61, and it remains to be seen how the NRC will interpret these regulations. As discussed in Chapter 4, "disposal site closure" is not a meaningful concept. For any storage/disposal methods, we see instead eternal vigilance. Any set-aside funds should make provisions for continual monitoring and repair for an indefinite period.

CONTACTS

If you would like to be active on radioactive waste issues, contact any of the local and national groups listed below.*

Alabama

James L. Taylor
Sierra Club—Alabama Chapter
P.O. Box 5591
Birmingham, AL 35205

Kyle Crider
Alabama Conservancy
2717 Seventh Ave S #201
Birmingham, AL 35233
205-322-3126

Larry Crenshaw
Safe Energy Alliance
P.O. Box 3241-A
Birmingham, AL 35255

Dr. Ed Passerini
New College
Box CD
University, AL 35486
205-683-8416

Alaska

John Hines/Jeff Bohman
Alaska PIRG
P.O. Box 1093
Anchorage, AK 99510
907-278-3661

Arizona

Jill Morrisson
Palo Verde Intervention Fund
6413 South 26th St
Phoenix, AZ 85040

Russel Lowes
Arizonians for a Better Environment
7501 East Hubbell
Scottsdale, AZ 85257
602-894-6314

Myron Scott
Coalition for Responsible Energy Ed
315 West Riviera Dr
Tempe, AZ 85282
602-968-2179

John Davis/Dave Foreman
Earth First!
P.O. Box 5871
Tucson, AZ 85703
602-622-1371

Arkansas

Bob Bland
Arkansas Alliance
8015 Brandon St.
Little Rock, AR 72204
501-565-3581

Glen White
Arkansas Sierra Club
4 Monica
Little Rock, AR 72204
501-562-6893/568-4294

Pat Costner Ex Dir
Environmental Congress of Ark
P.O. Box 548
Eureka Springs, AR 72632

Annee Littell
People's Action Safe Environment
322 Watson St
Fayetteville, AR 72701

California

Steve Aftergood
Committee to Bridge the Gap
1637 Butler Ave #203
Los Angeles, CA 90025
213-478-0829

Susan Birmingham
CalPIRG 1147 S Robertson Blvd
Los Angeles, CA 90035
213-278-9244

Andy Lieberman
Shut 'Em Down
10920 National Blvd #2
Los Angeles, CA 90064
213-470-1355

Edith Roth
San Fernando Valley Sierra Club
6029 Oakdale Ave.
Woodland Hills, CA 91367
818-346-9692

Liz Allen
Sierra Club—Angeles Chapt
394 E. Blaisdell Dr.
Claremont, CA 91711
714-624-5823

Eugene Cramer
Neighbors for the Environment
Box 9004
El Monte, CA 91733
818-964-1474

Jim Jacobson
Alliance for Survival
P.O. Box 33686
San Diego, CA 92103
619-277-0991

Dollie Irwin
Desert/Pass Action Group
420 N. Morongo Ave.
Banning, CA 92220

Carolyn Toenjes
Desert/Pass Action Group
1863 Park Drive
Palm Springs, CA 92262

P Savino/Dave Sabol
Chain Reaction
P.O. Box 1871
29 Palms, CA 92277
619-367-2491

Marion Pack/R.Hammel
Orange Cnty Alliance for Survival
200 N Main St #M-2
Santa Anna, CA 92701
714-547-6282

Director
People for a Nuclear Free Future
331 N Milpas St
Santa Barbara, CA 93103

Jim Dodson
Sierra Club
43730 N. Higbee Ave.
Lancaster, CA 93534

Ann McConnell
Citizens Against Yucca Dump
P.O. Box 693
Lone Pine, CA 93545
619-876-4682

Swift/Sharpe/Heried
Abalone Alliance Clearinghouse
2940 16th St Rm 310
San Francisco, CA 94103
415-861-0592

Gene Coan
Sierra Club
730 Polk Street
San Francisco, CA 94109
415-776-2211

Erika Rosenthal
Greenpeace
Fort Mason, Bldg. E
San Francisco, CA 94123
415-474-6767

Director
Natl Coalition to Stop Food Irradiatn
P.O. Box 590-488
San Francisco, CA 94159
415-566-CSFI

Mary Klein
Citizens for Altves to Nucl Power
P.O. Box 377
Palo Alto, CA 94302

Jean Sutherland
Sierra Club
3696 Bryant St.
Palo Alto, CA 94306
415-494-1277

Laurie Baumgarten
Nuclear Free California
1600 Woolsey St
Berkeley, CA 94703
415-841-6500

Director
Diablo Project Office
1530 1/2 Broad St
San Luis Obispo, CA 95401

Jim Adams/Michael Welch
Redwood Alliance
P.O. Box 293
Arcata, CA 95521
707-822-7884

Nini Redway
Sacramentans for Safe Energy
331 "J" St #105
Sacramento, CA 95814
916-444-SAFE

Madlyne MacDonald
League of Women Voters of Sacramento
2206 "K" St #2
Sacramento, CA 95816
916-443-3678

Colorado

Megan Seibel Dir
CoPIRG
1724 Gilpin St
Denver, CO 80218
303-355-1861

Director
Citzns Agnst Rocky Flats Contamn
1660 Lafayette
Denver, CO 80218
303-832-4508

Melinda Kassen
Environmental Defense Fund
1405 Arapahoe Ave
Boulder, CO 80302
303-440-4901

Amory Lovins
Rocky Mountain Institute
Drawer 248
Old Snowmass, CO 81654
303-927-3851

Connecticut

Judi Friedman
People's Action for Clean Energy
101 Lawton Road
Canton, CT 06019
203-693-4377

David Desirato
CT Citizen Action Group
51 Van Dyke Ave
Hartford, CT 06106

Chuck Morgan
People's Action for Clean Energy
123 Jordan Road RR1
Willimatic, CT 06226
203-423-5403

---

*Names were kindly provided by NIRS and Public Citizen and added to the Campaign's list of activists.   Any omissions were entirely unintentional. Please inform the Radioactive Waste Campaign of changes or if your group should be added to the list in the next edition.

Ed Mierzwinski
CT PIRG
U of Conn, Box U-8
Storrs, CT 06268
203-486-5002

Ralph Milione, Pres
Citizens Action Gp/Env Protctn
385 Main St #1-C
East Haven, CT 06512
203-469-2302

Delaware

Director
Nuclear Power Postponement
1612 East Robino Drive
Wilmington, DE 19808

District of Columbia

Kitty Tucker
Health and Energy Institute
236 Mass. Ave. N.E. #506
Washington, DC 20002
202-543-1070

Rob Hager/Lanny Sinkin
Christic Institute
1324 North Capitol St.
Washington, DC 20002
202-797-8106

Robert Alvarez/Fred Millar
Environmental Policy Institute
218 D St. S.E. 2nd Fl.
Washington, DC 20003
202-544-2600

Brooks Yaeger
Sierra Club
330 Pennsylvania Avenue, SE
Washington, DC 20003
202-547-1141

Kathleen Welch
US Public Interest Res.Group
215 Penn. Ave. SE Fl.3
Washington, DC 20003
202-546-9707

Ken Bossong
Public Citizen Critical Mass
215 Penn. Avenue, S.E.
Washington, DC 20003
202-546-4996

Sam White
Nuclear Waste Task Force
412 First St. S.E.#40
Washington, DC 20003
202-484-2773

Rick Parrish
Environmental Task Force
1012 14th St.NW 15th Fl
Washington, DC 20005
202-842-2222

Dan Reicher, Esq
NRDC
1350 NY Ave. NW #300
Washington, DC 20005
202-783-7800

Damon Moglen
Greenpeace
1611 Conn. Ave. NW
Washington, DC 20009
202-462-1177

Ellyn Weiss, Esq
Harmon & Weiss
2001 S St. N.W. #403
Washington, DC 20009
202-328-3500

G Coling/J Schwartz
Rural Coalition
2001 S St NW # 500
Washington, DC 20009

Greg and Brenda Johnson
Blacks Against Nukes
3728 NH Ave. N.W. #202
Washington, DC 20010
202-882-7155

Franklin Gage
Task Force/Nuclear Pollution
P.O. Box 1817
Washington, DC 20013
202-547-6661/474-8311

Ken Maize
Union Concerned Scientists
1616 P St. N.W. #310
Washington, DC 20036
202-332-0900

Diane D'Arrigo
NIRS
1616 P Street N.W. #160
Washington, DC 20036
202-342-9657

Tom Carpenter
GAP
1555 Conn.Ave. NW #202
Washington, DC 20036
202-232-8550

Scott Denman
SECC
1717 Mass.Ave.NW #LL215
Washington, DC 20036
202-483-8491

Caroline Petti
Southwest Research
2001 O Street N.W.
Washington, DC 20036
202-457-0545

Dave Culp/Chris Nichols
Environmental Action Foundation
1525 New Hampshire Ave. NW
Washington, DC 20036
202-745-4870

Florida

Joette Lorion
Center for Nucl Responsibility
7210 Red Road #217
Miami, FL 33143
305-661-2165

Jane Brown/Warren Hoskins
Conchshell Alliance
12040 S.W. 187 St
Miami, FL 33177
305-253-2635/576-3050

Geraldine Rasmussen
Citizens Against Rad Pollution
1609 S.E. Fourth St
Ft Lauderdale, FL 33301

Elise Jacques
Florida PIRG
1441 East Fletcher Ave
Tampa, FL 33612
813-971-7564

Georgia

Deborah Shepard
Campaign for Prosperous Georgia
1083 Austin Ave NE #107
Atlanta, GA 30307
404-659-5675/525-6458

Felix Rogers
Sierra Club Georgia
187 Degress Ave N.E.
Atlanta, GA 30307
404-522-9001

Jim Renner
Sierra Club
325 Southview Dr
Athens, GA 30605
404-546-8691

Hawaii

Edgar J.Y.Lin
U.S.-Taiwan Environmental Network
1127 11th Ave #207
Honolulu, HI 96826

Idaho

Liz Paul
Snake River Alliance
Box 4090
Ketchum, ID 83340
208-726-7728

Mary Kelly
Idaho Conservation League
P.O. Box 844
Boise, ID 83701
208-345-6933

Lisa Schultz
Snake River Alliance
P.O. Box 1731
Boise, ID 83701
208-344-9161

Wes/Gertie Hanson
Citizens Against Nucl Weapons
P.O. Box 2152
Coeur d'Alene, ID 83814
208-667-9389/8693

M Butters/C Brooscious
Clearwater Hanford Watch
P.O. Box 8582
Moscow, ID 83843
208-835-6125

J.R. Wilkinson
Hanford Educ Action League
Box 8654
Moscow, ID 83843
208-882-0167

Illinois

Lorens Tronet
Lake County Defenders
34 Maple Street
Crystal Lake, IL 60014
815-455-3243/642-9225

Mimi & Fred Targ
Citzns Opposed Rad Pollution
365 North Deere Park Rd
Highland Park, IL 60035
312-432-8247

Bruce von Zellen
De Kalb Area Alliance
P.O. Box 261
De Kalb, IL 60115
815-758-0829/756-2801

Dave Kraft
Nucl Energy Information Service
P.O. Box 1637
Evanston, IL 60204
312-869-7650/475-5696

Director
Safe Energy Now
910 Prairie St
Joliet, IL 60435

Howard Learner/D.Casel
Bus.& Prof. People in Public Interest
109 N Dearborn # 1300
Chicago, IL 60602
312-641-5570

Ed Gogol/Melody Moore
Citizens Against Nucl Power
220 South State #1202
Chicago, IL 60604
312-786-9041

N.Gardner/D Zalman
Illinois Safe Energy Alliance
53 W Jackson Blvd # 343
Chicago, IL 60604
312-663-1667

Joanne Hoelscher
Citizens for a Better Environment
33 E Congress St # 523
Chicago, IL 60605
312-939-1530/939-1984

Helen S. Lillibridge
Illinois Safe Energy Alliance
6285 N. Leona Avenue
Chicago, IL 60646
312-631-1307

R Rudner/Stan Cambell
Sinnissippi Alliance
219 E State St
Rockford, IL 61104
815-964-7111/963-0233

Betty Johnson
League of Women Voters
1907 Stratford Ln.
Rockford, IL 61107
815-399-0089

Gary McIntyre
Sierra Club-Blackhawk
1653 Fifth Ave.
Rockford, IL 61108
815-226-9779

Eloise Baker
Assoc. Cit. for Prot. of Env
P.O. Box 531
Sheffield, IL 61361
309-288-6501

Doris Conley
Sierra Club
3 Locust Hills Woods
Lebanon, IL 62254
618-537-2712

Indiana

Mick Harrison
INPIRG Activities Desk, Memorial Union
Indiana University
Bloomington, IN 47405
812-335-7575

Iowa

Jim Dubert
Iowa PIRG, Memorial Union Rm 36
Iowa State University
Ames, IA 50011
515-294-8094

Marty Hock
Citzns United for Responsible Energy
3500 Kingman Blvd
Des Moines, IA 50311
515-277-0253

Kansas

Stevi Stevens
Nuclear Awareness Network
1347 1/2 Massachusetts
Lawrence, KS 66044
913-749-1640

Bob Eye, Esq
Nuclear Awareness Network
1347 1/2 Mass Ave.
Lawrence, KS 66044

Claudia Spener
Kansas Natural Resources Council
1115 Horne
Topeka, KS 66604
913-357-6915/233-6707

Laura Menhusen
North Kansas Citizens Route 1
Jewell, KS 66949
913-428-3579

Dir—Nucl Program
Kansas Natural Resource Council
1516 Topeka Ave
Topeka, KS 66612
913-233-6707

Jolene Grabill
Kansans for Sensible Energy
1521 Fairview
Wichita, KS 67203

Director
Stop Nuclear Power
P.O. Box 8556
Wichita, KS 67208

Kentucky

Kathy Bond/Peg Dillinger
Paddlewheel Alliance
425 W Mohammed Ali Blvd
Louisville, KY 40202

Phyllis Fitzgerald
Environmental Alternatives
818 East Chestnut St
Louisville, KY 40204

Larry Wilson
Highlander Center
RR2   Box AA68
Middlesboro, KY 40965

Louisiana

Terri Karcher
Mothers for Safe Energy
2617 River Bend Dr
Violet, LA 70092

Carol Haire
Sierra Club—Nucl Power Committee
1435 Nashville Ave
New Orleans, LA 70115
504-899-4082/891-7086

Hal Dean
Sierra Club
1920 Dante St
New Orleans, LA 70118
504-861-1421

Russell Henderson
LA Citizens for Safe Energy
P.O. Box 15030
New Orleans, LA 70175

Philip H. Shafer
Acadiana Resource Conversion
101 Wilbourne Blvd. #503
Lafayette, LA 70506

Lesanne Kirkland
AWARE
400 Eden St
Plaquemine, LA 70764
504-687-6270

Ron/Virginia Martin
Sierra Club/Arklatex
380 Albany
Shreveport, LA 71105
318-861-7506

Maine

Susan Neilly
Lakes Environmental Association
102 Main St
Bridgeton, ME 04009
207-647-8580

Teri Steal/Bonnie Titcom
Citizens Against Nuclear Trash
P.O. Box 701
South Casco, ME 04077
207-655-4661/655-3706

Johansen/Stacy/Jones
Maine PIRG
92 Bedford St
Portland, ME 04102
207-780-4044

Carol Fritz
League of Women Voters/Maine
9 Hunt Club Woods
Cape Elizabeth, ME 04107
207-767-3737

Judy Barrows
Safe Power for Maine
P.O. Box 2204
Augusta, ME 04330
207-623-9231

Charlie Ipcar/Alva Morrison
Maine Nuclear Referendum Committee
Box 2627
Augusta, ME 04330
207-622-4395/772-2958

Ms. Sharon Treat
Natural Resource Council
271 State St
Augusta, ME 04330
207-622-3101

Bonnie Savage
Advisory Comm. on RadWaste
State House Station #120
Augusta, ME 04333
800-453-4013

Maria Holt
Safe Power for Maine
Box 115   High St
Bath, ME 04530

Nancy Chandler
Sierra Club
Box 253 Smallpoint Rd.
Sebasco Estates, ME 04565
207-389-1565

Maryland

Nuclear Program
Univ. of MD-MD PIRG
3110 Main Dining Hall
College Park, MD 20742
301-454-5601

Jean S. Ewing
Peach Bottom Alliance
3300 Jourdan
Darlington, MD 21034
301-457-4557

Pat Birnie MD
Nuclear Safety Coalition
6503 Overheart
Box 902 Columbia, MD 21045
301-730-0178/596-0924

Albert Donnay/Mairi MacRae
Nuclear Free America
325 East 25th St
Baltimore, MD 21218
301-235-3575

Tom Chalkley
Maryland Citizens Action
2500 N. Charles St
Baltimore, MD 21218

Massachusetts

Al Giordano
Mass Alert
52 Grinnell St
Greenfield, MA 01301
413-772-6098

Phil Stone
Central Mass Safe Energy
P.O. Box 1409
Worcester, MA 01601

Abigail Avery
Sierra Club
P.O.Box 246
Lincoln Center, MA 01773
617-259-8438

Thomas Moughan, Sr.
Citizens Within Ten-Mile Rad
Box 382
Amesbury, MA 01913
617-388-3640

Ingrid Sanborn
Citizens Within Ten-Mile Radius
85 Church St
West Newbury, MA 01985

Priscilla Chapman
Sierra Club
3 Joy St
Boston, MA 02108
617-227-5339

Rachel Shimshack
Massachusetts PIRG
29 Temple Place
Boston, MA 02111
617-292-4800

Amy Goldsmith
MA Nuclear Referendum Comm.
P.O. Box 1712
Boston, MA 02130
617-522-4970

Peter S. Turner
Greenpeace New England
139 Main Street
Cambridge, MA 02142
617-576-1650

Paul Smolens
Campaign for Safe Energy
53 Langley Road #210
Newton Center, MA 02159
617-964-4104

R. Reibstein
Radiation Events Monitor
162 Marrett Road
Lexington, MA 02173

Glenn Morrow
Mass Safe Energy Alliance
P.O. Box 1712
Boston, MA 02205
617-522-7196

Lois Traub
Union of Concerned Scientists
26 Church Street
Cambridge, MA 02238
617-547-5552

Gail Reed/Cheryl Nickerson
Pilgrim Alliance
98 Ellsville Road
Plymouth, MA 02360
617-224-3557

Michigan

Jennifer Puntenney
Safe Energy Coalition
33414 Oakland #3
Farmington, MI 48024
313-243-7974/477-3441

Brian Ewart
TOCSIN
2349 Arrowwood Trail
Ann Arbor, MI 48105,

Michael J. Keegan
Coalition Nuclear-Free Great Lakes
811 Harrison St
Monroe, MI 48161
313-241-6998

Diane Hebert
Greenpeace
2505 E. Sugnet
Midland, MI 48640

Mary Sinclair
Great Lakes Energy Alliance
5711 Summerset Dr
Midland, MI 48640
517-835-1303

Don Rounds   Dir
PIRGIM
220 N Chestnut St
Lansing, MI 48933
517-487-6001

LIVING WITHOUT LANDFILLS 91

Joanne Beemon
Conc Citizens of Charlevoix
204 Clinton
Charlevoix, MI 49720
616-547-4820

Minnesota

Mary Davis
Sierra Club Energy Committee
RR 3 Box 27
Northfield, MN 55057
507-645-6221

Sharon Collins
Northern Sun Alliance
1519 East Franklin Ave
Minneapolis, MN 55404
612-874-1540

Henry Fieldseth
Radwaste Transport Coalition
2412 University Ave SE
Minneapolis, MN 55414
612-376-7556/646-5632

Kristan Blake
Minnesota PIRG
2412 University Ave. SE
Minneapolis, MN 55414
612-627-4035

Barbara Johnson
Radioactive Waste Project
2412 University Ave SE
Minneapolis, MN 55414

Mississippi

Mark Tew
Sierra Club Box 817
University, MS 38677
601-234-4977

Stan Flint
Citizens Against Nucl Waste Disposal
3305 Old Mobile Highway
Pascagoula, MS 39567
601-762-7426

Missouri

Rich McClintock
MO PIRG
4144 Lindell Blvd # 410
St. Louis, MO 63108
314-534-7474

Dan Brogin
Crawdad Alliance
1420 South Big Bend
St. Louis, MO 63117
314-644-3014

Mary Weir
Coalition for the Environment
6267 Delmar Blvd.
St. Louis, MO 63130
314-727-0600

Kay & Leo Drey
Coalition for the Environment
515 West Point Ave
University City, MO 63130
314-725-7676

Diane Sheehan/Christa Wissler
CART
6267 Delmar Blvd
St. Louis, MO 63130
314-727-2311

Roger Pryor
Coalition for the Environment
6267 Delmar Blvd
St. Louis, MO 63130
314-727-0600

Montana

Russ Brown
Northern Plains Resource Council
419 Stapleton Building
Billings, MT 59101

Jim Jensen Ex Dir
Montana Environmental Info Center
P.O. Box 1184
Helena, MT 59624
406-443-2520

Brad Martin
MT PIRG 356 Corbin Hall
University of Montana
Missoula, MT 59802
406-721-6040

Nebraska

Marilyn D. McNabb
Nebraska for Peace
1836 H St Apt B
Lincoln, NE 68508
402-474-2964

Daryl LaPointe Sr.
Winnebago Tribe of Nebraska
PO Box 687
Winnebago, NE 68071

Ted Hoffman
Sierra Club
HC-74, Box 61-C
Chadron, NE 69337
308-432-5744

Sam Welsch
Western Nebraska Resources
Box 64
Crawford, NE 69339

Nevada

Bob Fulkerson
Citizen Alert
P.O. Box 5391
Reno, NV 89513
702-827-4200

New Hampshire

Kirk Stone
N H Energy Coalition
39 Old Sandown Road
Chester, NH 03036
603-887-4169

Director
NH Radioactive Waste Information
P.O. Box 193
Warner, NH 03278
603-847-9026

Paul Gunter/Ray Morrison
Clamshell Alliance
P.O. Box 734
Concord, NH 03301
603-224-4163/926-2686

Jackie Tuxill
Audubon Society of N.H.
Box 528-B Concord, NH 03301

Amena Anderson
N H Peoples Alliance
8 North Main
Concord, NH 03301
603-225-2097

Marjory Swope
NH Assoc. Conserv.Commissns
54 Portsmouth St.
Concord, NH 03301
603-224-7867

Cia Iselin
NE Coalition On Nucl.Pollution
Rte 1 Box 226
Marlborough, NH 03455
603-847-9026

Bob Backus/Jane Doughty
Seacoast Anti-Pollution League
5 Market St
Portsmouth, NH 03801
603-431-5089

Adam Auster
Seacoast/Clamshell/Seabrook Alliance
Box 1415
Portsmouth, NH 03801

Mary Scherer
C-10
P.O. Box 301
Exeter, NH 03833
603-772-6442

Carol Bellin
Clamshell Alliance
P.O. Box 877
Hampton, NH 03842
603-926-2686

New Jersey

Barbara Materna
NJ SEA Alliance
324 Bloomfield Ave
Montclair, NJ 07042

Linda Stansfield
League of Women Voters
71 Lake Dr.
Mountain Lakes, NJ 07046
201-334-1182

Larry Bogart
Citizens Energy Council
77 Homewood Ave
Allendale, NJ 07401
201-327-3914

Jeanne Oterson
New Jersey Citizen Action
380 Main St
Hackensack, NJ 07601

Responsible Management Rad Waste
17 Baird Pl
Whippany, NJ 07981

Diane Walker
Sierra Club, NJ Chapter
360 Nassau St.
Princeton, NJ 08540
609-924-3141

Peter Montague
Environmental Research Foundation
PO Box 3541
Princeton, NJ 08543

Ken Ward, Dir
NJ PIRG
84 Paterson St
New Brunswick, NJ 08901
201-247-4606

Director
NJ Environmental Lobby
46 Bayard St., Rm. 320
New Brunswick, NJ 08901
201-246-6832

New Mexico

Janet Greenwald
CARD
711 Arno S.E.
Albuquerque, NM 87102
505-242-7546

Lynda Taylor/Don Hancock
Southwest Research & Info Center
P.O. Box 4524
Albuquerque, NM 87106
505-262-1862

Gladys Winblad
Sierra Club/Rio Grande
6000 Buena Vista NW
Albuquerque, NM 87114
505-898-9762

David Benavides
New Mexico PIRG
Box 66 SUB/U New Mexico
Albuquerque, NM 87131
505-277-2757

New York

Joan Holt
NY PIRG
20 East 8th St 5th Fl
New York, NY 10003
212-477-6749

John Miller
Mobilization for Survival
853 Broadway Rm 418
New York, NY 10003
212-533-0008

Michael Zamm
Council on the Env. of NYC
51 Chambers St., Rm 228
New York, NY 10007
212-566-0990

Jennie Tichenor
Radioactive Waste Campaign
625 Broadway 2nd Fl
New York, NY 10012
212-473-7390

Madeline Dennis
NY-NJ Trail Conference
232 Madison Ave,Rm 908
New York, NY 10016
212-696-6800

Alan McGowan
Scientists Inst. for Public Info
355 Lexington Ave 16 Fl
New York, NY 10017

Joy McNulty
Nuclear Hazards Info Center
301 West 53rd St #10-K
New York, NY 10019

Jan Beyea
National Audubon Society
950 Third Ave
New York, NY 10022

Ken Gale
SHAD Alliance
P.O. Box 279
New York, NY 10185
212-249-6689

Robert Liebman
Sierra Club/Lower Hudson
70 Barker St. #206
Mt. Kisco, NY 10549
914-986-3437

Barbara Hickernell
Alliance to Close Indian Point
12 Terrich Court
Ossining, NY 10562
914-762-5922

Lindsay Audin
Citizens Against Nuclear Trucking
One Everett Ave
Ossining, NY 10562

Connie Hogarth
Westchester Peoples Action
255 Grove St
White Plains, NY 10602

Bernard Flicker
Rockland Families/Close Indian Point
27 Dogwood Ln N.
Pomona, NY 10970
914-362-1456

Jared Bark
WARD
35 Drew Road
Warwick, NY 10990
914-986-3437

Bill Weinberg
New York Greens
40 Park Pl
Brooklyn, NY 11217

Warren Liebold
Sierra Club, National Energy Comm
185 Glen Ave
Sea Cliff, NY 11579
516-674-3320

Dan Gluck/Murray Barbash
Citizens to Replace LILCO
137 Broadway
Amityville, NY 11701
516-691-5565

Director
PeaceSmith House
90 Pennsylvania Ave
Massapequa, NY 11758
516-798-0778

Nora Bredes
Shoreham Opponents Coalition
195 East Main St
Smithtown, NY 11787
516-360-3987

Anne Rabe
Safe Energy Coalition/EPL
196 Morton Ave
Albany, NY 12202
518-462-5526

Tom Ellis
Safe Energy Coalition of N.Y.
639 Myrtle Ave
Albany, NY 12208

Travis Plunkett
N.Y. PIRG
184 Washington Ave
Albany, NY 12210

Alex Cukan
Sierra Club
Emp State Pla, Box 2112
Albany, NY 12220
518-472-1534

Alfonso Scarpa
Tri-County Power Line Assoc.
RD-1, Box 193
Preston Hollow, NY 12469
518-239-4094

Anna Wasserbach
NY Federation fr Safe Energy
Box 2308, W. Saugerties Rd.
Saugerties, NY 12477
914-246-5700

Dick Hermans
Safe Energy Coalition of NYS
P.O. Box 495
Millerton, NY 12546

Shelagh Clancy
Syracuse Peace Council
924 Burnet Ave
Syracuse, NY 13203
315-472-5478

Amy Hubbard
Sierra Club/Iroquois Group
709 Ackerman Ave.
Syracuse, NY 13210
315-471-6447

Peter Kardas
Weapons Facilities Conv.Proj
821 Euclid Ave.
Syracuse, NY 13210
315-475-4822

Peter Blue Cloud
AKWESASNE NOTES
Mohawk Nation
Rooseveltown, NY 13683
518-358-9531

Carol Mongerson
Coalition on W Valley Nuclear Wastes
10734 Sharp St
East Concord, NY 14055
716-941-3168

Ray Vaughan
Coalition on W Valley Nuclear Wastes
135 East Main St
Hamburg, NY 14075

Charles Haynie
Buffalo Greens
93 Crescent Ave
Buffalo, NY 14214
716-837-6104

Elizabeth Thorndike
Center for Environ. Info.
33 S. Washington St.
Rochester, NY 14608
716-546-3796

Berta Hegeman
Coal. on WV Nuclear Wastes
141A Dutch Hill Rd.
Little Valley, NY 14755
716-938-6543

Theresa/David Fusco
Tritiums
RD #2
Bath, NY 14810
607-776-4555

North Carolina

Russell Norburn
NC Conservation Council
307 Granville Road
Chapel Hill, NC 27514
919-942-7935

Betsy Levitas
People's Alliance
212 N. Bloodworth St
Raleigh, NC 27601

Lisa Finaldi
Clean Water Fund of N.C.
P.O. Box 1008
Raleigh, NC 27602
919-832-7491

Nuclear Program
NC Radioactive Waste Watch
P.O. Box 11311
Raleigh, NC 27604
919-832-7491

Bill Holman
NC Conservation Council
1024 Washington St
Raleigh, NC 27605

Wells Eddelman
Alternatives Shearon Harris
812 Yancey St
Durham, NC 27701
919-942-1080/688-0076

Jesse Riley
Sierra Club
854 Henley Place
Charlotte, NC 28207
704-374-4342

Sandy Adair
Blue Ridge Envirnmntal Defense League
Rt. 3 Box 912
Boone, NC 28607
704-264-0259

Janet Hoyle
Blue Ridge Envirnmntal Defense League
P.O. Box 88
Glendale Spring, NC 28629
919-982-2691

Ron Lambe
Western NC Alliance
Rte 1 Box 127E
Bakersville, NC 28705
704-765-7187

Bill Crawford
Western NC Alliance
P.O. Box 1591
Franklin, NC 28734
704-524-3389

Paul Galimore/Kate Jayne
Long Branch Environmental Center
Rt. 2 Box 132
Leicester, NC 28748
704-683-3662

Lou Zeller
Madison Cty Envirnmntal Defense League
P.O. Box 291
Mars Hill, NC 28754
704-656-2773

Lindsay Jones
Western NC Alliance
Rt 1 Box 304, Green River Road
Zirconia, NC 28790
704-693-1702

Kitty Boniske
Carolinians For Safe Energy
Box 5855
Asheville, NC 28813
704-684-6680

Ohio

Mike Ferner
Toledo Coalition for Safe Energy
2975 113th St
Toledo, OH 43611
419-729-7273

Dini Schut
Toledo Coalition for Safe Energy
P.O. Box 4545
Toledo, OH 43620
419-536-6920/478-6284

Stephen Sass
Sunflower Alliance
1104 East 15th St
Ashtabula, OH 44004
216-964-3536

Deidre Francis
Conc Citizens of Geauga County
11210 Hidden Spring Dr
Chardon, OH 44024

Ronald O'Connell
Conc Citizens of Ashtabula County
315 Garfield St
Geneva, OH 44041

Susan L. Hiatt
Ohio Citizens for Responsible Energy
P.O. Box 22
Grand River, OH 44045
216-255-3158

Jim McIntyre
Sunflower Alliance
P.O. Box 151
Jefferson, OH 44047
216-576-5515

Gary Kalman
Ohio PIRG
Box 89 Wilder Hall
Oberlin, OH 44074
216-775-8137

Elbert J. Waldorf
Citizens Against Nucl Power
6470 Auburn Road
Painesville, OH 44077

Connie P. Kline
Conc Citizens of Lake County
38531 Dodds Landing Dr
Willghby Hill, OH 44094

D Sefcek/Bob Greenbaum
N Ohio Citizens Against Perry
14409 Bayes Ave
Lakewood, OH 44107
216-521-0567

LIVING WITHOUT LANDFILLS 93

Chris Trepal
Cuyahoga County Conc Citizens
1280 Manor Park
Lakewood, OH 44107
216-521-0004/843-7272

Rick Chudner
N.Ohio Citizens Against Perry/Davis
3715 West 139
Cleveland, OH 44111
216-941-8846

Bill Callahan
Ohio Public Interest Campaign
1501 Euclid Ave # 500
Cleveland, OH 44115
216-861-5200

Arnold Gleisser
Save Our State From Rad Waste
5505 South Barton Rd
Lyndhurst, OH 44124

Kim Hill
Sierra Club
5004 Oakland Rd
Lyndhurst, OH 44124
216-382-1853

Kathleen O'Neil
Sierra Club
709 Hemlock Dr
Euclid, OH 44132
216-241-7900/732-8518

Chris Borello
Concerned Cit. of Lake Twnshp
12018 Basswood Ave. NW
Uniontown, OH 44685
216-699-3224

Ned Ford
Sierra Club
6 Bella Vista Pl
Cincinnati, OH 45206
513-861-7807

Laura Yeamans
Ohio Public Interest Campaign
P.O. Box 2612
Athens, OH 45701

Oklahoma

Brian Hunt
Environmental Action
Box 38
Snow, OK 74567
405-298-2803

Jesse Deer-in-Water
Native Americans for Clean Environment
Rt. 2 Box 51-B
Vian, OK 74962
918-773-8184

Oregon

Lloyd Marbet
Forelaws On Board
19142 S Bakers Ferry Rd
Boring, OR 97009
503-637-3549

Timothy Schechtel
Columbia R. Fellowship for Peace
5565 Miller Rd.
Mt. Hood, OR 97041
503-352-6395

Eric Stachon
OSPIRG
027 S.W. Arthur St
Portland, OR 97201
503-222-9641

Betty McArdle
Oregon Environmental Council
2637 SW Water Ave
Portland, OR 97201
503-222-1963

Gregory Kafoury
Don't Waste Oregon Committee
320 SW Stark St #202
Portland, OR 97204
503-224-2647

Nina Bell/Eugene Rosalie
Coalition for Safe Power
408 SW 2nd Ave #406
Portland, OR 97204
503-295-0490

Joanne Oleksiak
Hanford Clearinghouse
921 SW Morrison Rm 420
Portland, OR 97205
503-295-2101

Drew Gardner
People Against Dumping at Hanford
P.O. Box 12051
Portland, OR 97212
503-224-3380

Pennsylvania

Cindee Virostek
Kiski Valley Coalition
409 N Eighth St
Apollo, PA 15613
412-478-2351

Judith Johnsrud
Environmental Coalition on Nucl. Power
433 Orlando Ave
State College, PA 16801
814-863-1972

D Hossler/James B Hurst
People Against Nucl Energy
P.O. Box 268
Middletown, PA 17057

E Epstein/Kay Pickering
TMI-Alert
315 Peffer St
Harrisburg, PA 17102
717-233-7897/3072

Jeff Schmidt
Sierra Club
P.O. Box 663
Harrisburg, PA 17108
717-232-0101

Jean Fix
York Environmental Alliance
2050 Deininger Rd
York, PA 17402

Frances Skolnick
Susquehanna Valley Alliance
P.O. Box 1012
Lancaster, PA 17604
717-872-7803

Phyllis Zitzer
Limerick Ecology Action,
Box 761,
Pottstown, PA 19464

Rhode Island

Dennis Roy
Rhode Island PIRG
228 Weybosset St 4th Fl
Providence, RI 02903
401-331-7474

South Carolina

Becky Hardee/Francie Hart
Energy Research Foundation
1916 Barnwell Street
Columbia, SC 29201
803-256-7298

Brett Bursey
GROW
18 Bluff Road
Columbia, SC 29201
803-254-4565/9398

F. Truett Nettles
Sierra Club/Lunz Group
7A Chadwick
Charleston, SC 29407
803-556-7704

Virginia Dykes
Sierra Club
207 Stone Ridge Rd.
Greer, SC 29651
803-268-7609

Judith Gordon
Sierra Club-Savannah River
c/o 150 Cypress Dr.
N. Augusta, SC 29841
803-279-4152

South Dakota

Jeanne Koster
S.D. Resources Coalition
RR #3, Box 253
Watertown, SD 57201
605-886-3532

Jim Macinnes
Sierra Club
5417 Twilight Drive
Rapid City, SD 57701
605-343-2901

Lawrence Perry
Black Hills Energy Coalition
P.O. Box 8092
Rapid City, SD 57701

Ray Lautenschlager
Fall River Cits v Nucl. Dump
Box 17
Ardmore, SD 57715
605-459-2671

Tennessee

Faith Young
Concerned Citizens of Tenn.
Rt. 1 Box 1 Hwy 25
Dixon Springs, TN 37057
615-832-0392

John Sherman
Tennessee Environmental Council
1719 West End Ave #227
Nashville, TN 37203
615-321-5075

Jenine Honicker
Church Women United
362 Binkley Drive
Nashville, TN 37211
615-832-0392

Ruth/Jack Neff
Sierra Club
2116 Westwood
Nashville, TN 37212
615-297-9870

Bill Leiper
Sierra Club-Tennessee
14000 Old Dayton Pike
Sale Creek, TN 37373
615-332-6671

Bob Pyle
Sierra Club
PO Box 16160
Chattanooga, TN 37416
615-899-3333

Susan Williams
SOCM
Box 457
Jacksboro, TN 37757
615-562-6247

Dir-Nuclear Program
Americans for a Clean Environment
P.O. Box 777
Kingston, TN 37763

David Twiggs
Americans for a Clean Environment
Rt.4 Lakeview Rd
Lenoir City, TN 37771
615-986-6899

Nettie Ballinger
S.I.C.K.
2625 Wilson Ave
Knoxville, TN 37996
615-524-8003

Leon Lowery
Tenn Valley Energy Coalition
1407 E. 5th Ave
Knoxville, TN 37917
615-637-6055

Jim Price
Sierra Club
S.E. Regional Office
P.O. Box 11248
Knoxville, TN 37939

Albert Bates Dir
The Natural Rights Center
156 Drakes Lane
Summertown, TN 38483
615-964-2334/3992

Louise Gorenflo
Tenn Nuclear Waste Taskforce
Rt. 6 Box 526
Crossville, TN 38555
615-788-2736

Texas

Bonnie Coatney
Armadillo Coalition of Texas
12130 Landlock
Dallas, TX 75218

Juanita Ellis
Citizens Assoc.for Sound Energy
1426 South Polk
Dallas, TX 75224

M Belisle/Jim Schrembeck
Comanche Peak Life Force
2710 Woodmere
Dallas, TX 75233
214-528-8792/421-1984

Gertrude Barnstone
Nuclear Safety League
1413 Westheimer
Houston, TX 77006

David Marrack
Houston Audubon Society
c/o 420 Mulberry Ln.
Bellaire, TX 77401

Pat Coy
Conc Citizens Against Nucl Power
5106 Caso Oro
San Antonio, TX 78233
512-653-0543

Rick Lowerre
Henry, Lowerre & Mason
2103 Rio Grande
Austin, TX 78705
512-479-8125

Dan Harrison
South Texas Cancellation Campaign
3400-B Lafayette
Austin, TX 78722
512-339-4844

Paul Robbins
South Texas Cancellation Campaign
Box 50484
Austin, TX 78763

Delbert Devin
Nuclear Waste Task Force
218 East Bedford
Dimmitt, TX 79027
806-647-5735/668-4678

Tonya Kleuskens
POWER
Route 1
Hereford, TX 79045
806-258-7583

Director
STAND
Route 2 Box 28
Tulia, TX 79088

George/Sharon Drain
STAND
6231 I-40 West #205
Amarillo, TX 79106
806-352-1662

Mrs. James Lynch
Alert Cit. for Env. Safety
Box 659
Dell City, TX 79837
915-964-2426

John Hamilton
Sierra Club
2708 Altura
El Paso, TX 79930

Utah

Steve Erickson
Don't Waste Utah Campaign
Box 1563
Salt Lake City, UT 84110

Vermont

Vt/N.H./Maine Green
N.E.Comm of Correspondence
P.O. Box 1342
White River Junction, VT 05001

Director
N.E.Coalition on Nuclear Pollution
P.O. Box 545
Brattleboro, VT 05301
802-257-0336

Diana Sidebotham/Esther Poneck
Hill and Dale Farms
R.D. 2, Box 1260
Putney, VT 05346
802-387-5817

Cort Richardson/Jackie Coates
VPIRG
43 State St
Montpelier, VT 05602
802-223-5221

Director
Citizens for Nuclear Free Vt
P.O. Box 53
Montpelier, VT 05602

Virginia

Lois Gibbs, Dir
Citizen's Clearinghouse
PO Box 926
Arlington, VA 22216
800-552-2075

Dave Levy
Sierra Club
Rt.7, Box 257
Charlottesville, VA 22901
804-978-1016

Donald Day
Sierra Club
151 Buckingham Circle
Charlottesville, VA 22903

Cory Thayer
Citizen Action Safe Energy
1502 London Co. Way
Williamsburg, VA 23185

Washington

Bill Mitchell
NW Nuclear Safety Campaign
2345 Franklin Ave #4
Seattle, WA 98102
206-324-2497

Hazel Wolf
Hanford Oversight Committee
512 Boylston Ave #106
Seattle, WA 98102
206-322-3041

Marc Sullivan
NW Conservation Action Coalition
3429 Fremont Pl #308
Seattle, WA 98103
206-547-6910

Jim Beard
Greenpeace/Good Shepherd Cntr
4649 Sunnyside Ave
North Seattle, WA 98103
206-632-4326

John Villani
League of Conservation Voters
1406 N.E. 50th St #201
Seattle, WA 98105
206-524-6554

L Notham/Wendy Wendlandt
WASHPIRG
340 15th Ave East
Seattle, WA 98112
206-322-9064

Tom Buchanan
NW Inlands Water Coalition
1438 20th Ave
Seattle, WA 98122
206-329-3839

Mark Reis
NW Conservation Action Coalition
Box 20458
Seattle, WA 98122

Barb Boyle/Annie Bringloe
Sierra Club
1516 Melrose Ave
Seattle, WA 98122
206-621-1696

Steve Rock
Southern Puget Sound SEA
1620 E Fourth St
Olympia, WA 98506

Nuclear Waste Program
Yakima Indian Nation
P.O. Box 151
Toppenish, WA 98948

Amy Mickelson
Physicians for Social Responsibility
S 325 Oak St
Spokane, WA 99204
509-624-7256

William Houff/Joan Mootry
Hanford Education Action League
S 325 Oak St
Spokane, WA 99204
509-448-8360/624-7256

West Virginia

Brian Gessner
West Virginia WVU PIRG
Mountainlair SOW
Morgantown, WV 26506
304-293-2108

Geoff Green
WV People's Energy Network
Rt. 1 Box 79-A
Burlington, WV 26710

Wisconsin

Peter Anderson
WI Second Decade
14 W Mifflin St #5
Madison, WI 53703
608-251-7020

Jane Elder
Sierra Club Midwest
214 N. Henry St., #203
Madison, WI 53703
608-257-4994

Director
Nuke Watch/Progressive Foundation
315 W Gorham St
Madison, WI 53703
608-256-4146

Joel Ario Dir
Wisconsin PIRG
520 University Ave
Madison, WI 53703
608-258-8200

Naomi Jacobson Dir
League Against Nuclear Danger
525 River Rd
Rudolph, WI 54475
715-423-7996

Director
Northwoods Alliance
P.O. Box 65
Tomahawk, WI 54487

Warren Viehl
Badger Alliance
North 6218 Jason St
Onalaski, WI 54650

Will Fantle
Badger Safe Energy Alliance
22 1/2 S Barstow St
Eau Claire, WI 54701

Director
Badger Safe Energy Alliance
P.O. Box 68
Durand, WI 54736
715-455-1099

Billie Garde
Govt Accountability Project
3424 North Marcos Lane
Appleton, WI 54911
414-730-8533

Canada

Rosalie Bertell
Inst of Concern for Public Health
67 Mowat Ave., Suite 343
Toronto, ONT, CANADA M6K-3E3
416-533-7351

Donovan Timmers
Concerned Citizens
15-764 Wolsely Ave.
Winnipeg, MAN, CANADA R3G-1C6

REFERENCES

ACP,1984	American College of Physicians, Health and Public Policy Committee, *Annals of Internal Medicine* 100, 1984, pp.912-913.
AIF,1986	Atomic Industrial Forum, "Building A Perspective," August 1986, Washington, D.C. The AIF has now merged with the U.S. Committee for Energy Awareness to form the U.S. Council for Energy Awareness.
Barlett,1985	Barlett, D.L. and J.B. Steele, *Forevermore*, W.W. Norton & Co., New York, 1985.
Battelle,1979	Battelle Columbus Laboratories, *Preliminary Environmental Implications of Alternatives for Decommissioning and Future Use of the Western New York Nuclear Services Center*, BMI-X698(Rev.), Columbus, Ohio, Feb 1979.
Baudisch,1985	Baudisch, H., et al, "DAW Volume Reduction Using the Newly Developed 2200 Ton Superpack—A New Generation of Supercompactor Equipment," in the *Proceedings of Waste Management '85, Vol. 2*, Tucson, Arizona, pp.513.
Blanchard,1987	Blanchard, Governor James, letter to Greg Larson, Director, Midwest Compact Commission, June 18, 1987.
Branagan,1986	Branagan, E.F. and F.J. Congel, "Disposal of Slightly Contaminated Radioactive Wastes from Nuclear Power Plants," presented at the meeting of the Health Physics Society, Knoxville, Tennessee, February 2-6, 1986.
Cahill,1983	Cahill, J., "Movement of Radionuclides in Unconsolidated Coastal Sediments at the Low-Level Radwaste Burial Site Near Barnwell, South Carolina," in *Proceedings of Waste Management '82*, 1983.
Carlson,1987	Carlson, E., "Quick, Name a State Willing to Accept Radioactive Waste," *The Wall Street Journal*, June 30, 1987, p.33.
Carter,1985	Carter, T.J. and P.K.M. Rao, "Fifteen Years of Radioactive Waste Management at Ontario Hydro," *Waste Management '85*, Tucson, Arizona, March 1985.
Cartwright,1982	Cartwright, K., "A Geological Case History: Lessons Learned at Sheffield, Illinois," Illinois State Geological Survey, Champaign, Illinois, 1982.
Cashman,1982	Cashman, T., New York State Department of Environmental Conservation, letter to Marvin Resnikoff, Sierra Club, Buffalo, NY, Feb 3, 1982.
Chinn,1984	Chinn, H., Office of the Illinois Attorney General, personal communication with Marvin Resnikoff, Sierra Club, Buffalo, N.Y., Dec 29, 1984.
Clymer,1986	Clymer, G., "10 CFR61 Irradiated Component Characterization at Crystal River Unit 3," in *Waste Management '86*, Tucson, Arizona, March 1986.
Cook,1984	Cook, J., et al, "Greater Confinement Disposal Program at the Savannah River Plant," *Proceedings of Waste Management '84*, Tucson, Ariz, March 11-15, 1984.
Cunningham,1985	R. Cunningham, Fuel Cycle Licensing Branch, NRC, personal communication with M. Resnikoff, Bethesda, Md., March 4, 1985.
Dames,1985	Dames and Moore, "Numbers and Types of Facilities Needed by 1986, 1991, and 1996," for the Southeast Compact Commission for Low-Level Radioactive Waste Management, July 1985.
Dayal,1983	Dayal, R., "Geochemical Investigations at Shallow Land Burial Sites," in *Proceedings of the Fifth Annual Participants' Information Meeting: DOE Low-Level Waste Management Program*, EG&G Idaho, Idaho Falls, Dec 1983.
DOE,1978	Department of Energy, *Western New York Nuclear Service Center Study, Companion Report*, TID-28905-2, Washington, D.C., Dec 1978.
DOE,1986	Department of Energy, *Characteristics and Inventories of Nuclear Waste*, Washington, D.C., Apr 1986.
DOH,1979	New York State Department of Health, News Release, Albany, N.Y., Apr 1979.
DOH,1986	New York State Department of Health, Data on Am-241 Levels in Sludges from the Tonawanda and Grand Island Sewage Treatment Plants..., K. Rimawi, Albany, New York, 1986.
Doyle,1985	Doyle, W., D.A. Drum, and J.D. Lauber, "The Smoldering Question of Hospital Wastes," *Pollution Engineering*, July 1985, pp. 35.
EGG,1983a	EG & G Idaho, Inc., *Managing Low-Level Radioactive Wastes: A Proposed Approach*, DOE/LLW-9, Idaho Falls, Apr 1983.
EGG,1983b	EG & G Idaho, Inc., *An Analysis of Low-Level Waste Disposal Facility and Transportation Costs*, DOE/LLW-6Td, Idaho Falls, Apr 1983.
EGG,1983c	EG & G Idaho, Inc., *1982 State-by-State Assessment of Low-Level Radioactive Wastes Shipped to Commercial Disposal Sites*, DOE/LLW-27T, Idaho Falls, Dec 1983.
EPA,1981	Environmental Protection Agency, *Federal Register*, February 5, 1981.

EPRI,1982	Electric Power Research Institute, *Identification of Radwaste Sources and Reduction Techniques*, EPRI NP-3370, Palo Alto, Cal., 1982.
Fisher,1983	Fischer, J.N., "U.S. Geological Survey Studies of Commercial Low-Level Radioactive Waste Disposal Sites—A Summary of Results," *Proceedings of the Fifth Annual Participants' Information Meeting: DOE Low-Program*, EG&G Idaho, Idaho Falls, Dec 1983.
GAO,1976	U.S. General Accounting Office, "Improvements Needed in the Land Disposal of Radioactive Wastes—A Problem of Centuries," RED-76-54, January 12, 1976.
Gay,1986	Gay, R.L. and L.F. Grantham, "High Strength Cementized Dried Resins," in *Waste Management '86*, Tucson, Arizona, March 1986.
Healey,1983	Healey, R.W., "Preliminary Results of a Study of the Unsaturated Zone at the Low Level Radioactive Waste Disposal Site Near Sheffield, Illinois," *Proceedings of the Fifth Annual Participants' Information Meeting: DOE Low-Level Waste Management Program*, EG&G Idaho, Idaho Falls, Dec 1983.
Helminski,1987	Helminski, E.L., "The Low-Level Radioactive Waste Policy Amendments Act of 1985, An Overview," *Waste Management '86*, Tucson, Arizona, March 1986.
Johnson,1983	Johnson, T.M., et al, "A Study of Trench Covers to Limit Infiltration at Waste Disposal Sites," *Proceedings of the Fifth Annual Participants' Information Meeting: DOE Low-Level Waste Management Program*, EG&G Idaho, Idaho Falls, Dec 1983.
Jur,1986	Jur, T.A., and W.M. Poplin, "A Critical Review of Materials Selected for High Integrity Containers," in *Waste Management '86*, Tucson, Arizona, March 1986.EG&G Idaho, Idaho Falls, Dec 1983.
Kentucky,1984	Kentucky Natural Resources and Environmental Protection Cabinet, *Comprehensive Low-Level Radioactive Waste Management Plan for the Commonwealth of Kentucky*, DOE/ID/12348-T6, Frankfort, Ky., Mar 1984.
Kriesberg,1986	Kriesberg, J., *Economic Benefits of a Nuclear Phaseout*, Public Citizen, Washington, D.C., Oct 1986.
Lipschutz,1980	Lipschutz, R.D., *Radioactive Waste: Politics, Technology and Risk*, Union of Concerned Scientists, Cambridge, Mass., 1980.
LLWG,1985	New York State Low Level Waste Group, lobbying brochure to the New York State Legislature, 1985.
Maine,1985	Maine Department of Environmental Protection, "The Siting, Design and Cost of Shallow Land Burial Facilities in Northern New England," Augusta, Maine, May 1985.
McConnell,1986	McConnell, J.W., and R.D. Saunders, "Degradation of Resins in Epicor-II Prefilters from Three Mile Island," in *Waste Management '86*, Tucson, Arizona, March 1986.
Meyer,1975	Meyer, G.L. and P.S. Berger, "Preliminary Data on the Occurrence of Transuranium Nuclides in the Environment at the Radioactive Waste Burial Site, Maxey Flats, Kentucky," International Symposium on Transuranium Nuclides in the Environment, IAEA, San Francisco, Nov 17-20, 1975.
Midwest,1986a	Midwest Interstate Low-Level Radioactive Waste Commission, "Regional Management Plan, Review of Alternative Waste Management Methods for the Midwest Compact Region," St. Paul, Minn., February 28, 1986.
Midwest,1986b	Midwest Interstate Low-Level Radioactive Waste Commission, "Regional Management Plan, Assessment of Impacts from Different Waste Treatment and Waste Disposal Technologies," St. Paul, Minn., May 23, 1986.
Mills,1983	Mills, D. and J. Razor, "An Infiltration Barrier Demonstration at Maxey Flats, Kentucky," *Proceedings of the Fifth Annual Participants' Information Meeting: DOE Low-Level Waste Management Program*, EG&G Idaho, Idaho Falls, Dec 1983.
Montague,1982	Montague, Peter, *New Jersey Hazardous Waste News*, Association of New Jersey Environmental Commissions, Mendham, New Jersey, 1982.
NEA,1981	Nuclear Energy Agency, Organization for Economic Cooperation and Development, "Cutting Techniques As Related To Decommissioning of Nuclear Facilities," Paris, France, February 1981.
Nogee,1986	Nogee, Alan, *Gambling for Gigabucks: Excess Capacity in the Electric Utility Industry*, Environmental Action Foundation, Washington, D.C., Dec 1986.
NRC,1978	Nuclear Regulatory Commission, *Technology, Safety and Costs of Decommissioning a Reference Pressurized Water Reactor Power Station*, NUREG/CR-0130, Washington, D.C., Jun 1978.
NRC,1980	Nuclear Regulatory Commission, *Technology, Safety and Costs of Decommissioning a Reference Low-Level Waste Burial Ground*, NUREG/CR-0570, Washington, D.C., Jun 1980.
NRC,1981a	Nuclear Regulatory Commission, *Evaluation of Trench Subsidence and Stabilization at Sheffield Low-Level Radioactive Waste Disposal Facility*, NUREG/CR-2101, Washington, D.C., May 1981.
NRC,1981b	Title 10 (Energy), Part 61 of the Code of Federal Regulations, *Licensing Requirements for Land Disposal of Radioactive Waste, Federal Register*, Vol. 46, pp. 38081, July 24, 1981.
NRC,1981c	Nuclear Regulatory Commission, 10 CFR Part 20, *Federal Register*, Vol 46, pp.16230, March 11, 1981.
NRC,1981d	Nuclear Regulatory Commission, *Draft Environmental Impact Statement on 10 CFR Part 61*, "Licensing

	Requirements for Land Disposal of Radioactive Waste," NUREG-0782, Washington, D.C.,Sep 1981.
NRC,1982a	Nuclear Regulatory Commission, Licensing Requirements for Land Disposal of Radioactive Waste, 10 CFR Parts 2, 19, 20, 21, 30, 40, 51, 61, 70, 73, and 170, *Federal Register,* Vol. 47, pp. 57446, December 27, 1982.
NRC,1982b	Nuclear Regulatory Commission, *Final Environmental Impact Statement on 10 CFR Part 61*, "Licensing Requirements for Land Disposal of Radioactive Waste," NUREG-0945, Washington, D.C., Nov 1982.
NRC,1982c	Nuclear Regulatory Commission, "Environmental Impact Appraisal and Safety Evaluation Report of Low-Level Radioactive Waste Storage at Tennessee Valley Authority, Sequoyah Nuclear Plant," Docket No. 30-19101, Washington, D.C., Sep 1982.
NRC,1983	Nuclear Regulatory Commission, *Characterization of the Radioactive Large Quantity Waste Package of the Union Carbide Corporation*, NUREG/CR-2870, prepared by Brookhaven National Laboratory, Washington, D.C., November 1983.
NRC,1984a	Nuclear Regulatory Commission, Memorandum to All Agreement States Regarding Sewer Sludge Contamination, from D. A. Nussbaumer, Office of State Programs, Washington, D.C., 1984.
NRC,1984b	Nuclear Regulatory Commission, *Proceedings of the Workshop on Shallow Land Burial and Alternative Disposal Concepts*, NUREG/CP-0055, Bethesda, Md., May 2-3, 1984.
NRC,1984c	Nuclear Regulatory Commission, *Scoping Study of the Alternatives for Managing Wastes Containing Chelating Decontamination Chemicals*, NUREG/CR-2721, Washington, DC, Feb 1984.
NRC,1984d	Nuclear Regulatory Commission, *Geologic and Hydrologic Research at the Western New York Nuclear Service Center, West Valley, New York*, NUREG/CR-3782, Washington, DC, Jun 1984.
NRC,1985a	Nuclear Regulatory Commission, Office of Inspection and Enforcement, IE Information Notice No. 85-57, Washington, D.C., July 16, 1985.
NRC,1985b	Nuclear Regulatory Commission, *Extended Storage of Low-Level Radioactive Waste: Potential Problem Areas*, NUREG/CR-4062, prepared by Brookhaven National Laboratory, Upton, N.Y., Dec 1985.
NRC,1986a	Nuclear Regulatory Commission, *Update of Part 61 Impacts Analysis Methodology*, by Envirosphere Co., NUREG/CR-4370, Washington, D.C., January 1986.
NRC,1986b	Nuclear Regulatory Commission, *Technical Considerations Affecting Preparation of IX Resins for Disposal*, NUREG/CR-4601, prepared by Brookhaven National Laboratory, May 1986.
NRC,1986c	Nuclear Regulatory Commission, Below Regulatory Concern, 10 CFR Part 2, *Federal Register*, Vol. 51, pp. 30839, August 29, 1986.
NRC,1986d	Nuclear Regulatory Commission, *Preliminary Assessment of the Performance of Concrete as a Structural Material for Alternative Low Level Radioactive Waste Disposal Technologies*, NUREG/CR-4714, prepared by Brookhaven National Laboratory, Dec 1986.
NRC,1987a	Nuclear Regulatory Commission, Definition of High-Level Radioactive Waste, 10 CFR Part 60, *Federal Register*, Vol. 52, pp.5992, February 27, 1987.
NRC,1987b	Nuclear Regulatory Commission, "Map of LLRW Compact States," Office of State Programs, Washington, D.C., Feb 1987.
NRC,1987c	Letter from Knapp, M.R., Office of Nuclear Material Safety and Safeguards, Nuclear Regulatory Commission, to M. Resnikoff, Radioactive Waste Campaign, June 2, 1987.
Nucleonics,1987	Nucleonics Week, "Texas Becomes First State to Allow Landfill Burial of Some LLW," pp.1, June 18, 1987.
NUS,1980	NUS Corporation, *The 1979 State-by-State Assessment of Low-Level Radioactive Wastes Shipped to Commercial Burial Grounds*, NUS-3440, San Francisco, Nov 1980.
O'Donnell,1983	O'Donnell, E., "Insights Gained from NRC Research Investigations at the Maxey Flats LLW SLB Facility," *Proceedings of the Fifth Annual Participants' Information Meeting: DOE Low-Level Waste Management*, EG&G Idaho, Idaho Falls, Dec 1983.
ORNL,1984	Oak Ridge National Laboratory, *Plan for Diagnosing the Solvent Contamination at the West Valley Facility Disposal Area*, ORNL-2416, Oct 1984.
OTA,1985	Office of Technology Assessment, *Superfund Strategy, Summary,* OTA-ITE-253, Washington, D.C., Mar 1985.
Resnikoff,1977	Resnikoff, M., "Sierra Club Testimony Related to Section IV E, Reprocessing, Final GESMO I, March 4, 1977," NRC Docket No. RM-50-5, Washington, D.C., Mar 1977.
Rickard,1983	Rickard, W.H. and Kirby, L.J., "Tritium in a Deciduous Forest Adjacent to a Commercial Shallow Land Burial Site: Implications for Monitoring to Detect Radionuclide Migration," *Proceedings of the Fifth Annual Participants' Information Meeting: DOE Low-Level Waste Management Program*, EG&G Idaho, Idaho Falls, Dec, 1983
Russell,1987	Russell, R., "Fear of What Might Be,"*Topeka Capital-Journal*, Topeka, Kansas, April 15, 1987.
RWC,1985a	Radioactive Waste Campaign, "Burning Radioactive Waste: Is it Safe? Can it Work?," New York, May 1985.

RWC,1985b	Radioactive Waste Campaign, "'Low-Level' Nuclear Waste: Options for Storage," New York, 1985.
RWC,1986	Radioactive Waste Campaign, "Burning Radioactive Waste: What Comes Out of the Stack?", New York, Mar 1986.
Science,1984	"Juarez: An Unprecedented Radiation Accident," *Science*, March 16, 1984, pp.1152.
Shapiro,1981	Shapiro, Fred, *Radwaste*, Random House, New York, 1981
Siskind,1986	Siskind, B., *et al*, "Extended Storage of Low-Level Radioactive Waste: Potential Problem Areas," in *Proceedings of Waste Management '86*, Tucson, Arizona, Mar 1986.
Stegman,1987	Stegeman, Mark, "Low-Level Radioactive Waste Management in North Carolina: An Economic Analysis of Policy Options," for the Clean Water Fund of North Carolina, Raleigh, May 1987.
Sullivan,1986	Sullivan, W., "Solar Energy Gets a Boost from Flurry of Designs," *The New York Times*, New York City, April 29, 1986, p.C-1.
Tschirley,1986	Tschirley, Fred, "Dioxin," *Scientific American*, Feb 1986, pp. 29.
Union,1986	Union Electric, *Final Safety Analysis Report for the Callaway Nuclear Plant*, St. Louis, Missouri, Jun 1986.
Voice,1984	New York Voice of Energy, Letter to the New York State Assembly, February 6, 1984.
White,1987	White, A., letter to C Gordon, *et al*, Amherst, N.H., February 22, 1987.

Index

Index

above ground storage of wastes
  advantages and disadvantages of, 67-69
  buildings for, 64
  costs of, 72-74
  human intrusion in, 69-70
  vaults for, 64-67
absorbed liquids, 18-19
accidents, at nuclear power reactors, 11
activation products, 14
alternatives to landfills
  above-ground systems as, 64-70
  conservation as, 59-60
  economics of, 72-74
  below ground, 70-72
  separation by half-life as, 62-64
  storage or disposal as, 64
  volume reduction as, 60-62
American College of Physicians, 5
americium, 5
americium-241, 5, 22, 60
Amersham (firm), 20
Arkansas Nuclear One reactor, 11
Atomic Energy Commission (AEC), 33
Atomic Industrial Forum, 4
augered holes, 71

Babcock & Wilcox (firm), 19
Barnwell (South Carolina), 33, 34, 40-41
"bathtub effect," 34, 35
Beatty (Nevada), 33-35, 41
below ground storage of wastes, 70-72
"below regulatory concern" (BRC) class of wastes, 51-52, 59, 61, 79-80
biowaste, 18
  incineration of, 61
Blanchard, James, 53
boiling water reactors
  total wastes from, 26
  volume of wastes from, 16
boric acid, 11
buildings, above ground storage of wastes in, 64
bunkers, earth-mounded, 70-71
burning wastes, 60-62
  institutional, 19

California Nuclear Company, 36
capital costs, 74
carbon-14
  burning, 19
  industrial, 20
  in institutional wastes, 18
cartridge filters, 12
cesium, 5
cesium-137, 12, 17, 22
chelating agents
  EDTA, 35
  at landfills, 34
  NRC regulation 10 CFR Part 61 on, 47, 51
  at West Valley, 38, 39
Chem-Nuclear Systems (firm), 40
Cintichem (firm), 22
Class A wastes
  incineration of, 61
  NRC regulation 10 CFR Part 61 on, 47-48
Class B wastes
  earth-mounded bunkers for, 70
  NRC regulation 10 CFR Part 61 on, 47, 48
Class C wastes
  deep underground repositories for, 72
  earth-mounded bunkers for, 70
  47-49, 51
  reclassified as "low-level" wastes, 52
cleanups of landfills, 41-42
closure costs, 74
coal mines, 71
cobalt-60, 14, 50
compaction of wastes, 60-61
  from nuclear power reactors, 12
  institutional, 19
concentrated liquids, from nuclear power reactors, 11
concrete
  in earth-mounded bunkers, 70-71
  for storage vaults, 65-66
conservation of radioactive wastes, 59-60, 78
containers, see waste containers
costs
  under Low-Level Radioactive Waste Policy Act (1980), 53
  recommendations on, 80
  of waste disposal, 60

  of waste management techniques, 72-74
  see also economic issues
curium-242, 51

Dartmouth College, 64
decommissioning
  costs of, 60
  landfills, 36, 38
  nuclear power reactors, wastes from, 13-17, 28, 49
deep underground repositories, 72
dioxin, 19, 61-62
disposal of radioactive wastes, 64
  above ground, 64-70
  below ground, 70-72
Duke Power (firm), 52
dumping at sea, 33

earth-mounded bunkers, 70-71
economic issues
  in decommissioning landfills, 60
  in incineration, 62
  jobs affected by "low-level" wastes, 4
  in Low-Level Radioactive Waste Policy Act (1980), 53, 55
  in waste management, 72-74
  see also costs
EDTA (chelating agent), 35
electric generation, 60
Energy, U.S. Department of, 79
  wastes created by, 5
environmental effects, 55

federal government
  under Low-Level Radioactive Waste Policy Act (1980), 53
  see also Nuclear Regulatory Commission
filter sludges, 12
fission, 10
  NRC regulation 10 CFR Part 60, on high-level wastes from, 22
food irradiation, 5, 60
France, bunkers used in, 70-71
fuel, in nuclear power reactors, 10-12, 60

Galpin, Floyd, 52
gases
   from nuclear power reactors, 12
   produced by incineration, 61
geology of landfills, 33-34
group 4 waste streams, 49

half-lives
   in measuring hazards, 25
   separation of wastes by, 62-64, 78
   of tritium, 20, 21
hazardous life of nuclear wastes, 12, 55
hazards, NRC regulation 10 CFR Part 20 on, 25
health effects, 55
high-activity generators, 21
High Integrity Containers (HICs; waste containers), 11
high-level wastes
   deep underground repositories for, 72
   definition of, 9n
   military, 5
   not covered by NRC regulation 10 CFR Part 61, 24-25
   NRC regulation 10 CFR Part 20 on, 63
   NRC regulation 10 CFR Part 60 on, 22
   from nuclear power reactors, 10
   reclassified as "low-level" wastes, 52, 79
human intrusion, in extended storage systems, 69-70
hydrogen gas explosions, 67

Idaho Falls (Idaho), 33
incineration, 60-62
   of institutional wastes, 19
industrial wastes, 9, 19-24
institutional wastes, 5, 18-19
iodine, 61
iodine-129, 12
   hazardous life of, 26
ion exchange resins, 11
   hazardous life of, 26
   liquids in, 67
   Quadricells for, 66
iridium-129, 50

jobs affected by "low-level" wastes, 4
Juarez (Mexico), 1984 incident in, 50, 67
Jur, T.A., 69

Kansas, 3

laboratory wastes, 19
landfills, 41-42
   at Barnwell, 40-41
   chelating agents in, 15-16
   costs of, 72-74
   at Maxey Flats, 35-36
   NRC regulation 10 CFR Part 61 on, 25, 45-49
   problems at, 33-35
   at Richland and at Beatty, 41
   search for new sites for, 77-78
   at Sheffield, 36-37
   at West Valley, 37-40
   see also alternatives to landfills
leakages from landfills
   above-ground storage leaks compared with, 67-68
   covered by NRC regulation 10 CFR Part 61, 25
legislation, 45
   amendments of 1985 to Low-Level Radioactive Waste Policy Act (1980), 53-55
   decision-making and, 79
   Low-Level Radioactive Waste Policy Act (1980), 52-53, 79
   see also regulations
liabilities, under Low-Level Radioactive Waste Policy Act (1980), 53
licensing requirements, 53
lifespan of nuclear power reactors, 14
limestone mines, 71
liquid scintillation devices, 18
liquid wastes
   compaction not possible with, 61
   NRC regulation 10 CFR Part 60, on high-level wastes in, 22
   NRC regulation 10 CFR Part 61 on, 47
   from nuclear power reactors, 11
   in storage vaults, 65
Love Canal (New York), 50
Low-Level Radioactive Waste Policy Act (1980), 45, 52-53, 79
   amendments of 1985 to, 53-55
"low-level" wastes
   from decommissioned nuclear power reactors, 15
   definition of, 9
   hazardous life of, 24-26
   high-level wastes reclassified as, 52
   industrial, 19-24
   institutional, 18-19
   NRC regulation 10 CFR Part 61 on, 45-49, 51
   radioactivity from, 26
   recommendations for, 78-79
   sources of, 4-5, 27-28

Maine, 74
Manhattan Project, 50
Manhattan Project II, 78, 80
Maxey Flats (Kentucky), 33, 35-36, 72
maximum permissible concentration (MPC), 25-26
McGuire Air Force Base (New Jersey), 3
medical "low-level" wastes, 5, 18
Menhusen, Laura, 3
metal mines, 71
military high-level wastes, 5
mined cavities, 71-72
Minnesota Mining and Manufacturing (3M; firm), 22
module storage systems, 66-67
molybdenum-99, 22
Montague, Peter, 42
Montclair (New Jersey), 3
Morocco, 1984 incident in, 50

New England Nuclear (firm), 20
New Jersey, 3
New York State Low-Level Waste Group, 4
New York Voice of Energy, 4
nickel-59, 13, 14
niobium-94, 13, 14
non-fuel wastes, from nuclear power reactors, 12-13
North Central Kansas Citizens, 3
Nuclear Engineering Company (NECO), 35
   see also U.S. Ecology
Nuclear Fuel Services (NFS; firm), 37-38, 62
Nuclear Metals, Inc., 23
nuclear power industry, 55
   NRC regulations and, 51
   warnings on "low-level" wastes

104 LIVING WITHOUT LANDFILLS

from, 4-5
nuclear power reactors, 17-18
   conservation of wastes at, 59-60
   as containers for wastes, 74
   decommissioning, wastes from 13-17, 28
   fuel-related wastes from, 10-12
   non-fuel wastes from, 12-13
   total wastes from, 26-27
   wastes from, 5
   wastes stored at, 78
Nuclear Regulatory Commission (NRC)
   on extended storage, 69
   on ion exchange resins, 11
   regulation 10 CFR Part 20, on hazards, by, 25, 63
   regulation 10 CFR Part 60, on high-level wastes, by, 22
   regulation 10 CFR Part 61, on waste management, by, 22, 24-25, 45-49
   regulations adopted by, 51
   "below regulatory concern" (BRC) class of wastes under, 79-80
   scenarios used by, 50
   on sludges and contaminated materials, 12
   Update of Part 61 Impacts Analysis Methodology by, 5
nuclear submarines, wastes from, 23
Nuclear Waste Policy Bill (1980), 45
nuclear weapons, by-products of production of
   radioisotopes as, 22
   tritium as, 22
   wastes created by, 5

Oak Ridge National Laboratory (Tennessee), 5, 22, 33
occupational hazards, 69
operating costs, 74

Pharoah's Tomb scenario, 50
phosphorus-32, 18
plutonium
   at landfills, 35
   reprocessing of, 37
   in wastes created by nuclear weapons production, 5
   at West Valley, 39
plutonium-238, 51
plutonium-241, 22
polyvinyl chloride plastics, 61-62

Poplin, W.M., 69
pre-operational costs, 74
pressurized water reactors, 16-17
   total wastes from, 26
property values, 55
public participation, 55

Quadrex Corporation, 18
Quadricells, 66

radioactive animal carcasses, 18
radioactive wastes, types of, 9n
radioactivity
   from boiling water reactors, 16
   from "low-level" wastes, 26
   over 100 to 10,000 years, 28
   reduction of, 59
   from tritium, 21
radioisotopes, 9
   industrial, 19, 22
radionuclides
   in institutional wastes, 18
   NRC regulation 10 CFR Part 61 on, 47-49, 51
   regulations of hazards from storage of, 25
   in sealed sources, 22
   separation by half-lives of, 63
reactor sites, radioactive wastes at, 28-29, 74
recycling of radioactive materials, 67
regulations
   adopted by NRC, 51
   "below regulatory concern" (BRC) class of wastes under, 61, 79-80
   legislation and, 52-53
   NRC 10 CFR Part 20, on hazards, 25, 63
   NRC 10 CFR Part 60, on high-level wastes, 22
   NRC 10 CFR Part 61, on waste management, 22, 24-25, 45-49
   scenarios used for, 50
repositories, deep underground, 72
reprocessing
   wastes from, 24
   West Valley plant for, 37-38
Richland (Washington), 33-35, 41
Rocky Flats (Colorado), 22

salt mines, 71
Savannah River Plant (South

Carolina), 22, 71
scintillation liquids, 61
sea dumping, 33
sealed-sources generators, 21-22
separation of wastes by half-lives, 62-64, 78
Sequoyah Nuclear Plant, 64-65
sewer systems, institutional wastes in, 19
Sheffield (Illinois), 33, 36-37
sludge
   institutional wastes in, 19
   NRC on, 52
smoke detectors, 5, 22, 60
states, 45
   decision-making by, 79
   in Low-Level Radioactive Waste Policy Act (1980), 52-54
storage of radioactive wastes, 64
   above ground, 64-70
   below ground, 70-72
Stringfellow Acid Pits (California), 42
strontium-90, 12, 17, 35
submarines, wastes from, 23
supercompactors, 60-61
Surepak system (Westinghouse), 66-67

tailings, 9n
technetium-99m, 22
Tennessee Valley Authority (TVA), 64-66
Three Mile Island nuclear power reactors (Pennsylvania), 11
tools, radioactive, 50
transuranic waste, 9n, 10
   NRC regulations on, 51
trash, from nuclear power reactors, 12
trees, tritium in, 35-36
trenches and trench caps
   at Barnwell, Richland and Beatty, 41
   "bathtub effect" in, 34
   at Maxey Flats, 35
   at Sheffield, 36, 37
   at West Valley, 38
tributyl-n-phosphate (TBP), 39
tritium
   at Barnwell, 41
   burning, 19
   as by-product of nuclear weapons production, 5
   in institutional wastes, 18
   manufacturing wastes involving,

LIVING WITHOUT LANDFILLS 105

20-21
  at Maxey Flats, 35-36
  released by incineration, 61
  at Sheffield, 36-37

underground repositories, 72
uranium
  fission of, 10
  industrial wastes from, 23
  tailings, 9n
uranium-235, 10
U.S. Ecology (firm), 19, 35, 36

Vaults, above ground storage of wastes in, 64-67
Vernon (New Jersey), 3
Virginia Electric Company, 64
volume reduction, 60-62, 78

warehouses, above ground storage of wastes in, 64
waste containers, 11, 34
  in above-ground storage, 68
  NRC on, 69
  NRC regulation 10 CFR Part 61 on, 47
waste management
  conservation in, 60
  economic issues in, 72-74
  NRC regulation 10 CFR Part 61 on, 22, 24-25, 45-49
Waste Management, Inc., 40
waste streams, 5
  group 4, 49
  industrial, 23
  in nuclear power reactors, 10
  separation by half-lives of, 63, 78

water
  broken down by radiation, 67
  groundwater, contamination of, 42
  in nuclear power reactors, 11
  in waste containers, 11
water infiltration at landfills
  at Barnwell, Richland and Beatty, 41
  at Maxey Flats, 35
  at West Valley, 38
Westinghouse (firm), 36
  Surepak system by, 66-67
West Valley (New York), 23-24, 33, 37-40
  leachate overflow in, 69
  radioactive tools taken from, 50
  water in trenches in, 69

Radioactive Waste Campaign

The Radioactive Waste Campaign is a non-profit public interest organization based in New York City. It promotes greater public awareness of the dangers to human health and the biosphere from the generation of radioactive waste. The Campaign conducts a wide range of research, information dissemination and public education activities, and it produces books, fact sheets, slide shows, videos and a quarterly newspaper on radioactive waste issues, *the Waste Paper.*

BOARD OF DIRECTORS

Lisa Finaldi, Chair* Chip Hoagland
Carol Mongerson, Vice-Chair* Cia Iselin
Betty Quick, Secretary* Warren Liebold
Mike Cohn, Treasurer* June Peoples
Abigail Avery Dave Pyles
Jed Bark Simeon Sahaydachny
Priscilla Chapman Jeff Schmidt

Member, Executive Committee

STAFF

Minard Hamilton, Director Marvin Resnikoff, Research Director
Steven Becker, Editor Alan Spatafore, Office Assistant
Ed Hedemann, Production Director Jennie Tichenor, Office Manager & Bookkeeper

ADVISORY BOARD

Robert Alvarez, Environmental Policy Institute
Sr. Rosalie Bertell, Ph.D., Institute of Concern for Public Health
Jackson Davis, Ph.D., Univ. of California at Santa Cruz
Jessie Deer-in-Water, Native Americans for a Clean Environment
Kay Drey, Coalition for the Environment
Lois Gibbs, Citizen's Clearinghouse for Hazardous Wastes
Judith Johnsrud, Ph.D., Environmental Coalition on Nuclear Power
Charles Komanoff, Komanoff Energy Associates
John Mohawk, Daybreak
Robert Pohl, Ph.D., Cornell University
Bonnie Raitt, musician
Alice Stewart, Ph.D., Queen Elizabeth Medical Centre, England
Sting, musician
Ellyn Weiss, Esq., Harmon & Weiss

INTERNS

Mike Alfieri, Univ. of Vermont Law School Barry Seidman, Rutgers University
Robin Biggs, Phillips-Andover Academy Stephanie Wallace, Madison High School

Radioactive Waste Campaign Publications

FACT SHEETS

Insecure Landfills: The West Valley Experience
Detailed discussion of the solid radioactive waste dump, the so-called "low-level" waste burial ground at West Valley. Exposes the underground migration of water into the burial trenches. Seven lessons and warnings of that experience that can be utilized nationwide. 8 p.

"Low-Level" Nuclear Waste: Options for Storage
With the current push to reopen old burial grounds and site new dumps across the U.S., this fact sheet explores other storage methods for radioactive waste. Experiences in Canada, Tennessee and New Hampshire are discussed. 8 p.

Radioactive Waste: Buried Forever?
Profiles the six commercial radioactive landfills citing generic problems and history of radioactive leakage with brief discussion of alternatives. 6 p.

Incineration: What Is Coming Out of the Stack?
Assesses the radiological and other health hazards of incinerating radioactive wastes and considers types of waste burned, dioxins, accidents and alternatives. 8 p.

SLIDE SHOW

Hidden Legacy - A Profile of Radioactive Burial Grounds
Review of the poor record of "low-level" waste dumps across the country, specifically at Maxey Flats, Ky.; Sheffield, Ill.; and West Valley, N.Y. An important organizing tool for activists, especially those confronted by regional "low-level" waste compacts. Updated 1986. Snappy soundtrack.

NEWSLETTER

The Waste Paper
The quarterly newspaper of the Radioactive Waste Campaign, the world's first newspaper to specialize in radioactive waste issues. Not a rehash of the news you've seen elsewhere, but hard-hitting, original reporting. The latest news on breakthroughs or failures in waste technology. Up-to-date reports on citizen battles all over the country. Tips on resources and organizing. Only $8 per year for this important quarterly.

Support our work and keep up with the latest developments on the radioactive waste front. Become a member of the Radioactive Waste Campaign for $20 per year. Membership includes a discount on books and reports, action updates, and a subscription to the *Waste Paper.* To become a member or to order the above publications, write or call the Radioactive Waste Campaign.

Radioactive Waste Campaign • 625 Broadway, 2nd Floor • New York, New York 10012 • (212) 473-7390

BIOGRAPHIES

Marvin Resnikoff is Research Director of the Radioactive Waste Campaign. He received a Ph.D. in high energy theoretical physics from the University of Michigan in 1965, and has been on the Campaign staff since 1978. For the period 1981-1983, he was a Project Director at the Council on Economic Priorities where he authored the book, *The Next Nuclear Gamble*. Dr. Resnikoff has been a technical consultant on nuclear waste matters to the States of New York, Illinois, Utah and Kansas, the State of Lower Saxony, West Germany, and numerous local and national environmental organizations. Since 1974, he has testified many times before the U.S. Congress and State Legislatures on nuclear fuel reprocessing, waste management and transportation. Prior to his work at the Campaign, he taught at Rachel Carson College and the Department of Physics at the State University of New York at Buffalo.

Steven Becker is the Radioactive Waste Campaign's Editor. He holds an M.A. in political science from Columbia University, where he conducted research on energy issues. Prior to coming to the Campaign, he co-edited the international human rights bulletin *Peace and Democracy News*. Mr. Becker's previous work has included serving on the American Public Health Association's Task Force on Occupational and Environmental Disease, and working at the President's Council on Environmental Quality.

Ed Hedemann is manager of production and design for the Radioactive Waste Campaign's publications. He holds a B.A. in astronomy from the University of California (Berkeley) and spent four years in graduate school at the University of Texas (Austin). In the early 1970's, Mr. Hedemann was an instructor of astronomy at the University of Texas Night School. From 1973 to 1986 he was an organizer on the national staff of the War Resisters League in New York City. Ed is the author and editor of *Guide To War Tax Resistance,* and editor of the *War Resisters League Organizer's Manual*. He also wrote and produced several other publications, such as the "High School Organizing Packet" and the "ROTC Dismantling Kit."